T0332386

NANOSTRUCTURAL BIOCERAMICS

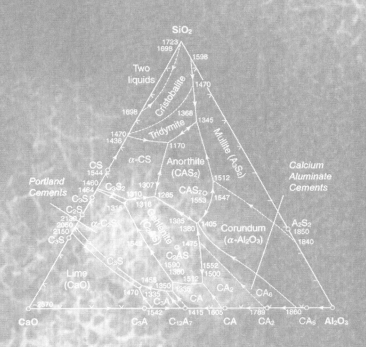

NANOSTRUCTURAL BIOCERAMICS

Advances in Chemically Bonded Ceramics

Leif Hermansson

PAN STANFORD PUBLISHING

Published by

Pan Stanford Publishing Pte. Ltd.
Penthouse Level, Suntec Tower 3
8 Temasek Boulevard
Singapore 038988

Email: editorial@panstanford.com
Web: www.panstanford.com

British Library Cataloguing-in-Publication Data
A catalogue record for this book is available from the British Library.

ISBN   978-981-4463-43-0 (Hardcover)
ISBN   978-981-4463-44-7 (eBook)

Printed in the USA

Contents

Preface

It is a great honor to present this book on 'nanostructural chemically bonded bioceramics'. The direct opportunity of this opened up after a speech on 'Why even difficult to avoid nanostructures in chemically bonded calcium aluminate–based biomaterials'. The invitation to write this book from Pan Stanford Publishing is thankfully acknowledged.

Writing a book, which covers a whole new technology within biomaterials science (materials, processing, properties, biological response, clinical evaluation, new applications, etc.), is of course not one person's work. I would like to acknowledge the following people contributing to the thinking in this book. I will start with my wife Irmeli, a dental technician, who challenged the author some decades ago with the question 'Why don't you from your ceramic platform do something which makes sense?' She wanted a substitute for amalgam. This started a work at Karolinska Institute at the former Center for Dental Technology and Biomaterials, Stockholm University, Sweden. Prof. Rune Söremark, late Associate Prof. Folke Sundström, and Associate Prof. Yangio Li are specifically acknowledged. The input from CEO Torgny Nilsson of KRISS, Sweden, was fundamental for the start of the new activities. After some turbulent years, the author met his 'positive anti-picture,' Dan Markusson, who has been of great general help in understanding biomaterials product development. Dan is now CEO of Peptonic Medical AB, Sweden. The work at Karolinska Institute, and later at Uppsala University, Sweden, have contributed enormously to the understanding of nanostructural chemically bonded biomaterials. The work by Prof. Håkan Enqvist, Tech. Drs. Lars Kraft and Jesper Lööf, and Associate Prof. Erik Adolfsson are thankfully acknowledged. Early cooperation with Prof. Roger Carlsson and Associated Prof. Elis Carlström and colleagues at Swedish Ceramic Institute (now within IVF-SWEREA), Prof. Richard Bradt, Pennsylvania State University, USA, and Prof. Hans Larker at former ABB Cerama AB (now Saint Gobain Advanced Ceramics AB), Sweden, have been fundamental for basic understanding of materials science. The author would like to thank all personnel within Doxa AB,

Sweden, and people related to development activities with several universities in Sweden and Europe. Finally I would like to express my great thank for support from relatives and friends in different ways.

Leif Hermansson
Summer 2014

Chapter 1

Introduction to Nanostructural Chemically Bonded Bioceramics

This chapter gives an introduction to chemically bonded bioceramics (CBBCs) and the position of CBBCs in relation to other biomaterials based on metals, polymers, and other bioceramics. A detailed description of CBBCs will be presented in the subsequent chapters.

1.1 Chemically Bonded Bioceramics: An Overview

Biomaterials in general are based on the materials' groups' metals, polymers, and ceramics [1]. Typical metallic biomaterials are based on stainless steel, cobalt-based alloys, and titanium or titanium alloys and amalgam alloys. Polymeric biomaterial composites from monomers are based on amides, ethylene, propylene, styrene, methacrylates, and/or methyl methacrylates. Biomaterials based on ceramics are found within all the classical ceramic families: traditional ceramics, special ceramics, glasses, glass-ceramics, coatings, and chemically bonded ceramics (CBCs) [2, 3]. Examples are given in Table 1.1.

Nanostructural Bioceramics: Advances in Chemically Bonded Ceramics
Leif Hermansson
Copyright © 2015 Pan Stanford Publishing Pte. Ltd.
ISBN 978-981-4463-43-0 (Hardcover), 978-981-4463-44-7 (eBook)
www.panstanford.com

Table 1.1   Examples of bioceramics

Ceramics: classification	Examples of bioceramics
Traditional ceramics	Dental porcelain, leucite-based ceramics
Special ceramics	Al, Zr, and Ti oxides
Glass	Bioglass ($Na_2O–CaO–P_2O_5–SiO_2$)
Glass ceramics	Apatite–wollastonite, Li-silicate-based
CBCs	Phosphates, aluminates, silicates, and sulphates

CBCs are widely used as general construction materials but have found new applications as biomaterials. The following CBC systems have been proposed or are already used as biomaterials: Ca-phosphates, Ca-silicates, Ca-aluminates, Ca-sulphates, and Ca-carbonates (see Table 1.2).

Table 1.2   CBBC systems

Group/name	Basic system
Calcium silicates	CSH[a]
Calcium aluminates	CAH[b]
Calcium phosphates	CPH[c]
Calcium sulphates	$CaSO_4–H_2O$
Calcium carbonates	$CaO–CO_2$

[a]CSH = $CaO–SiO_2–H_2O$ (calcium silicate hydrate).
[b]CAH = $CaO–Al_2O_3–H_2O$ (calcium aluminate hydrate).
[c]CPH = $CaO–P_2O_5–H_2O$ (calcium phosphate hydrate).

Most ceramics are formed at high temperatures through a sintering process. By using chemical reactions, the biomaterials in the chemically bonded bioceramic (CBBC) systems can be produced at low temperatures (body temperature), which is attractive from several perspectives: cost, avoidance of temperature gradients, (thermal stress), dimensional stability, and minimal negative effect on the system with which the material interacts. Notable is that the hard tissue of bone and teeth (apatite, a Ca-phosphate-based material) also is formed via a biological chemical reaction and close in composition to some of the CBBCs. The proposed CBBC systems have in general favourable biocompatible properties. The chemistry of these systems is similar to that of the hard tissue found in living organisms. These are based on different types of apatites and carbonates. The bioactivity aspects of CBBCs will be treated separately in Chapters 3 and 8.

One of the first ceramics to be proposed as a biomaterial was gypsum, $CaSO_4 \cdot \frac{1}{2}H_2O$. The first cement to be proposed and used was a Zn-phosphate, which is still used as a dental cement. Examples of typical phases formed in CBC systems are presented in Table 1.3.

Table 1.3 CBC systems

Group/name	Basic system	Typical phases formed
OPC[a]	CSH	Amorhous CSH, tobermorite
CAC[b]	CAH	Katoite and gibbsite
Gypsum plaster	Ca-sulphates	$CaSO_4 2H_2O$
Sorel	$MgO-H_2O$ (Cl)	MgOCl
Bioglasses	$CaO-Na_2O-SiO_2-P_2O_5$	Carbohydroxyapatite
Phosphates	CPH	Apaties, brushite, monetite
Carbonates	$CaO-CO_2-H_2O$	Calcite, aragonite
Geopolymers[c]	Aluminosilicates, metakaolin	Amorphous phases

[a]OPC = ordinary Portland cement
[b]CAC = high-alumina cement: C_3A, $C_{12}A_7$, CA, etc.
[c]Geopolymers = metakaolin or synthetic aluminosilicates

CBCs constitute ceramics which are being formed due to chemical reactions. Often the precursor material is a ceramic powder (e.g., Ca-silicate or Ca-aluminate), which is 'activated' in a water-based liquid. A chemical reaction takes place, in which the initial powder is partly or completely dissolved and new phases precipitate. The precipitated phases are composed of species from both the liquid and the precursor powder. The precipitates can be formed in situ, in vivo, often in the nanoscale due to low solubility of the phases formed (see Chapter 6 for details). The nanostructural CBBCs are especially found within the Ca-phosphate, Ca-aluminate, and Ca-silicate systems. The large pores between the original dissolving precursor powders are filled, and the material hardens. The dissolution speed and solubility products of the formed hydrate phases determine the nanosize, setting time, and final curing (hardening) of the material. The setting time can be controlled by selection of the precursor grain size and/or by addition of accelerating or retarding substances. Since the material can be formed from a precursor powder mixed with a liquid, the material can be made mouldable simply by controlling the amount of liquid (in relation to the powder) and by the possible addition of small amounts of polymers in the liquid. This makes CBBCs useful as injectable biomaterials, where the final biomaterial is formed in situ, in vivo. This will be treated in detail in Chapter 3 and Chapters 6–8. Also worth mentioning is a relatively new group of ceramics called geopolymers [4–6]. These are also produced by chemical reactions but do not involve hydration, that is, new uptake of water. The geopolymerisation is thus not a hydration process, in which water is consumed. Instead the water resides in the pores but plays an active role as a dissolution medium during the reaction, an inorganic polymerisation.

CBBCs can further be divided into two main groups, resorbable and stable biomaterials. This will be described next.

1.2 Stable and Resorbable Chemically Bonded Ceramics

Ca-aluminate-based biomaterials and to some extent Ca-silicates are stable materials after hydration and can favourably be used for load-bearing applications. The Ca-phosphates, Ca-sulphates, and Ca-carbonates are known to be resorbable or slowly resorbable when inserted in the body, and their main applications are within bone void-filling with low mechanical stress upon the biomaterial. The resorbable materials are after various times depending on the specific chemical composition replaced by new bone tissue. This can start immediately after injection, and the material can be completely dissolved after months and in some cases after a few years [7].

1.2.1 Stable Chemically Bonded Bioceramics

The chemistry and phases in stable CBBCs are presented in Fig. 1.1 [8]. The actual phases, using the cement abbreviation system ($C=CaO$, $A=Al_2O_3$, $S=SiO_2$, $H=H_2O$, etc. [9]) in the Ca-aluminate system are C_3A, $C_{12}A_7$, CA, and CA_2 and in the Ca-silicate system are C_3S and C_2S. The reactivity of all these phases increases with the content of CaO. When these phases are used as biomaterials, precaution must be taken in the hydration reactions[, which are exothermic reactions. The temperature increase during the setting and initial hardening can be controlled by selection of the phases and volumes involved and by processing agents.

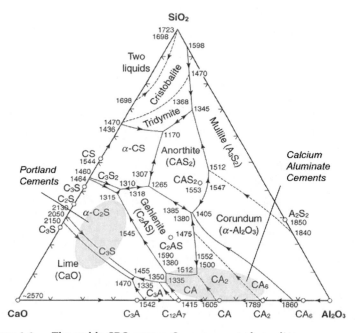

Figure 1.1   The stable CBC system: 3-component phase diagram.

The Ca-aluminate and Ca-silicate reaction mechanisms differ in one respect considerably from those of Ca-phosphate reactions and other reactions of resorbable chemically bonded materials. This has to do with the amount of water involved in the hydration reactions, and this can easily be seen when comparing the chemical formula of the phases formed in the curing reactions (see Table 1.4). An example of the water involved in the hydration of a given reaction is shown here:

$$3CaO \cdot Al_2O_3 + 12\ H_2O \rightarrow Ca_3[Al(OH)_4]_2(OH)_4 + 4Al(OH)_3$$

   Ca-aluminate   Water   Katoite   Gibbsite

The water content of the final chemical products has great influence upon the optimisation of the water-to-cement (w/c) ratio with regard to handling properties and the final microstructure, especially the residual porosity, which can be reduced substantially for the systems where the water consumption is high. The inherent high water uptake of stable CBBCs in comparison to resorbable ones yields benefits such as higher mechanical strength; extended possibility to add fillers, for example, for improved radio-opacity; and tuneable handling properties.

Table 1.4 The water involved in the hydration of three CBBC systems [2]

System	Typical phase(s)	Oxide formula	Mol. % H_2O	Wt.% in hydrated products
Ca-phosphate	Apatite	$10CaO \cdot 3P_2O_5 \cdot H_2O$	7	Approx. 5
Ca-aluminate	Katoite + gibbsite	$3CaO \cdot Al_2O_3 \cdot 6H_2O$ + $Al_2O_3 \cdot 2H_2O$	>60	Approx. 25
Ca-silicate	Tobermorite + amorphous phases	$5CaO \cdot 6SiO_2 \cdot 5H_2O$ + Ca, Si)H_2O	>30	Approx. 20

A typical hydrated area in Ca-aluminate is shown in Fig. 1.2. High-resolution transmission electron microscopy (HRTEM) reveals nanosize porosity <3 nm, often 1–2 nm, and hydrates in the size range of 20–50 nm [10, 11] (see Fig. 1.3).

1.2.2 Resorbable Chemically Bonded Bioceramics

Ceramic biomaterials are often based on phosphate-containing soluble glasses and various calcium phosphate salts. These materials can be made to cure in vivo and react in situ with the surrounding tissue and are attractive as replacements for the natural calcium phosphates of mineralised tissues. Ca-phosphate products are gaining ground in orthopaedics as resorbable bone substitutes. Other resorbable ceramics are based on Ca-sulphates or combinations of Ca-sulphates and Ca-phosphates. So far most bioceramic applications have dealt with bone replacement using phosphates. The scientific work and literature deal almost extensively with Ca-phosphate-based biomaterials [12].

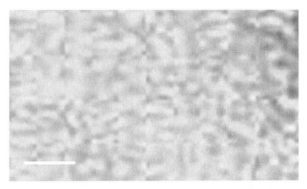

Figure 1.2 The hydrated area of a Ca-aluminate-based material (white bar = 50 nm).

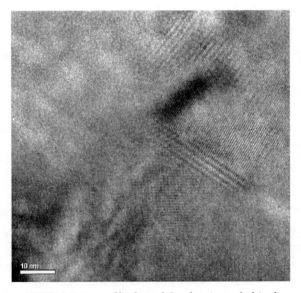

Figure 1.3 HRTEM picture of hydrated Ca-aluminate (white bar = 10 nm).

These materials are often designed in combination with implants with porosity or channels, into which the resorbable materials are introduced. The resorbable materials are supposed to be exchanged by new bone tissue. Therefore their mechanical properties are not constant and change over time. Resorbable as well as stable CBCs can also be used as coatings on implants materials (metals or ceramics) for permanent use or as replacements.

Table 1.5 summarises the most frequently used resorbable CBBC systems [13, 14].

Table 1.5   Examples of resorbable CBBC systems and main phases

System	Initial phases	Final phase
Ca-sulphate	a- $CaSO_4 \cdot 1/2H_2O$	$CaSO_4 \cdot 2H_2O$ (gypsum)
Ca-sulphate	b- $CaSO_4 \cdot 1/2H_2O$	$CaSO_4 \cdot 2H_2O$ (gypsum)
Ca-phosphate	b-TCP + mono-Ca-phosphate	$CaHPO_4$ (monetite)
Ca-phosphate	Solution of Ca^{2+} and PO_4^{3-}	$Ca_5(PO_4)_3(OH)$ (apatite)
Ca-phosphate	a-tricalcium phosphate	$(CaHPO_4)2H_2O$ (brushite)

1.3   Summary and Conclusions

- Most ceramics are formed at high temperatures through a sintering process. By using chemical reactions, CBBCs can be produced at low temperatures (body temperature), in situ, in vivo.
- The chemistry of these systems is similar to that of the hard tissue found in living organisms. These are based on different types of apatites and carbonates. CBBCs easily form nanostructures with crystal sizes similar to those found in hard tissue.
- Both stable and resorbable CBBCs can be produced. The stable phases are found within the CAH and CSH systems, while resorbable phases are seen within the CPH system and within sulphate systems.

  In summary CBCs have advantages related to many other biomaterials, including sintered bioceramics, with respect to:
  - the possibility of in situ–, in vivo–formed biomaterials;
  - reaction temperatures close to those of tissues;
  - similarity in chemistry and physics to hard tissue;
  - tolerance for moist conditions;
  - closure of contact zones;

- nanostructural integration; and
- bone and dental void–filling possibilities.

Acknowledgement

The author thanks the personnel at Doxa Company, Sweden, and the Materials Science Department at Uppsala University for valuable inputs during a two-decade period.

References

1. Park, J. B., and Lakes, R. S. (2007), *Biomaterials: An Introduction* (Springer, New York).

2. Hermansson, L. (2011), Nanostructual chemically bonded Ca-aluminate based biomaterials, in *Biomaterials: Physics and Chemistry*, Ed. R. Pignatello (INTECH, Rijeka).

3. Park, J. B. (2008), *Bioceramics* (Springer, New York).

4. Davidovits, J. (2008), *Geoplymer: Chemistry & Applications* (Institut Géopolmère, Saint Quentin, France).

5. Duxson, P., Fernández-Jiménez, A., Provis, J. L., Lukey, G. C., Palomo, A., van Deventer, J. S. J. (2007), Geopolymer technology: the current state of the art, *J. Mater. Sci.,* **42**(9), pp. 2917–2933.

6. Forsgren, J. (2010), PhD thesis, Uppsala University (Acta Universitatis Upsaliensis).

7. Hench, L. (1998), Biomaterials: a forecast for the future, *Biomaterials,* **19**, pp. 1419–1423.

8. Muan, A., and Osbourne, E. A., Ed. (1965), *Phase Equilibria among Oxides* (Addison-Wesley, New York).

9. Mangabhai, R. J. (1990), *Calcium Aluminate Cement, Conference Proceedings* (Chapman and Hall, London).

10. Hermansson, L., Lööf, J., Jarmar, T. (2008), Injectable ceramics as biomaterials: today and tomorrow, *Proc. ICC 2*, Verona.

11. Engqvist, H., Schultz-Walz, J. E., Loof, J., Botton, G. A., Mayer, D., Phaneuf, M. W., Ahnfelt, N.-O., Hermansson, L. (2004), Chemical and biological integration of a mouldable bioactive ceramic material capable of forming apatite in vivo in teeth, *Biomaterials,* **25**, pp. 2781–2787.

12. Ravagliolo, A., and Krajewski, A. (1992), *Bioceramics* (Chapman and Hall, London).

13. Nilsson, M. (2003), *Injectable Calcium Sulphates and Calcium Phosphates as Bone Substitutes*, PhD thesis, Lund University (Chapman and Hall, London).

14. Åberg, J. (2011), *Premixed Acidic Calcium Phosphate Cements*, PhD thesis, Uppsala University (Acta Universitatis Upsaliensis, Uppsala).

Chapter 2

Structures of Hard Tissue and the Importance of in situ–, in vivo–Formed Bioceramics

This chapter presents the structure of hard tissues—enamel, dentine, and bone tissue—and how chemically bonded bioceramics (CBBCs) interact with hard tissues. The features of CBBCs as injectable in situ–, in vivo–formed biomaterials will be treated in some detail. A detailed description of CBBC materials will be presented in Chapters 3 and 6–8.

2.1 Hard Body Tissue Structures: An Overview

Hard tissue is basically divided into three major groups: enamel, dentine, and bone structures [compact and spongy (trabecular or cancelleous)] [1]. The main purpose of hard tissue is to carry loads or have the ability to withstand mechanical pressure or stress. For the dental tissues, enamel and dentine, resistance against wear and chemical attack are also important features. The hard chemical component in all hard tissues is apatite. Apatite can appear in nature and in living beings in several modifications, as the basic structure hydroxyapatite (HA), $Ca_5(PO_4)_3(OH)$ easily forms solid solutions [2]. Examples are in Table 2.1.

Nanostructural Bioceramics: Advances in Chemically Bonded Ceramics
Leif Hermansson
Copyright © 2015 Pan Stanford Publishing Pte. Ltd.
ISBN 978-981-4463-43-0 (Hardcover), 978-981-4463-44-7 (eBook)
www.panstanford.com

Table 2.1   Examples of solid solutions of HA as minerals and biological apatites

Type of apatite	Main formula	Major substituent(s)	Comments
HA	$Ca_5(PO_4)_3(OH)$	–	Basic structure
F-aptite	$Ca_5(PO_4)_3(OH,F)$	F	Mineral
Mineral OH-apatite	$Ca_5(PO_4)_3(OH)$	–	Mineral
Dahllite	$Ca_5(PO_4, CO_3)_x(OH)_y$	CO_3	Mineral
Staffelite	$Ca_5(PO_4,CO_3)_x(OH, F)_y$	CO_3, F	Mineral
Human enamel apatite	$(Ca,Mg)_x(PO_4,HPO_4,CO_3)_y(OH,Cl)$	Mg, HPO_4, CO_3, Cl	Biological apatite
Shark enameloid	$Ca_x(PO_4,CO_3, HPO_4)_y(OH, F)$	F, CO_3, and HPO_4	Biological apatite

The united cell of apatite is $Ca_{10}(PO_4)_6(OH)_2$, which is why solid solutions in Table 2.1 can be described as $(Ca, Mg)_{10}(PO_4, CO_3, HPO_4)_6(OH, Cl, F)_2$. Other substituents in low concentrations are also possible, especially other metals such as Sr, Ba, Na, Li, Mn, and Zn. Thus three main positions—one for cations and two for anions— are found within the apatite structure. In many chemically bonded bioceramics (CBBCs) apatite structures are formed. In several other CBBCs two of these three positions are the same as in apatite. As an example for hydrated Ca-aluminate the phase $Ca_3[Al(OH)_4]_2(OH)_4$ is shown.

The amount of apatite in enamel is very high, approximately 96 wt.%. In dentine and bone tissues the apatite content is approximately 60% and 35%, respectively. The soft tissue part in hard structures is different types of collagen, intracellular matrix, and water.

The structures of apatite in hard tissue are designed to meet requirements on the macro-, micro-, and nanosize levels with regard to the formation and mechanical properties developed. All structures are based on nanosize crystals and nanosize inter- and intralayers [1, 3]. See Table 2.2.

Table 2.2   Tooth structure sizes

Tooth structure	Size in mm	Size in mm	Size in nm	Size in nm
Whole tooth	10	–	–	–
Individual plates	–	0.2	–	–
Nanocrystals	–	–	15	–
Single crystal within a nanocrystal	–	–	–	1.5

The importance of the nanostructure of biological apatite material and the nanostructure of CBBCs will be dealt within section 2.2.1 in this chapter and further in Chapters 6–8.

2.2   Interaction between Chemically Bonded Ceramics and Hard Tissue

All hard tissue structures are sensitive to temperature changes. Artificial materials such as biomaterials can practically only be placed in vivo and formed in situ if the temperature and temperature increase can be controlled at low levels.

Biomaterials which can be formed at body temperatures belong to some of the following material groups: solid solutions of soft or fluid metals (e.g., gold and mercury, amalgams), organic polymers formed by condensation (e.g., resins) or cross-linking (e.g., glass ionomer cement), and chemically bonded ceramics (e.g., Ca-aluminate- and Ca-silicate-based materials, Ca- and Zn-phosphates, and Ca-sulphates). All the CBBC materials have a similarity in chemistry to that of apatite, and these biomaterials can be injected into the hard tissue structure and formed in situ, in vivo, mainly due to the reaction pattern involving hydration mechanisms. Water reacts with the original phase(s), and hydrate(s) are formed.

The material when introduced in the body as a dental or as an orthopaedic material is in most cases a paste.

2.2.1   Contact Zone Reaction between Chemically Bonded Bioceramics and Hard Tissue

The chemistry, including the phases formed, and the structures obtained with CBBCs in contact with hard tissue facilitate and improve the connection between the biomaterial and the biological tissue. In many cases a nanostructural integration occurs. Five or six reaction mechanisms have been identified, which all contribute to a safe contact zone, chemically and physically [4]. These reactions are summarised in Table 2.3. More details concerning the mechanisms are dealt with in Chapters 3 and 6–8.

Table 2.3 Summary of chemical reactions facilitating integration between chemically bonded ceramics and hard tissue

Reaction	Feature
Mechanism 1	Main hydration reactions (aluminates, silicates, phosphates, or sulphates)
Mechanism 2	Apatite formation in the presence of phosphate ions in the biomaterial (aluminates and silicates)
Mechanism 3	Apatite formation in the contact zone in the presence of body liquid (aluminates, silicates)
Mechanism 4	Transformation of hydrated phases into apatite
Mechanism 5	Biologically induced integration and ingrowth
Mechanism 6	Mass increase reaction when unhydrated Ca-aluminate or Ca-silicate is present

The main reaction involves precipitation of nanocrystals on contact areas and within the injected material. Repeated precipitation occurs until the original powder or the water is consumed, resulting in closing of cavities, gaps, and voids. This opens up for multipurpose use as a biomaterial in many different applications, depending on the selection of the chemically bonded ceramic system. This is presented in detail in Chapters 9–11.

Complementary reactions occur when the Ca-aluminate or Ca-silicate is in contact with tissue containing body liquid. Several mechanisms have been identified which control how the material is integrated into tissue. These mechanisms affect the integration differently depending on what type of tissue the biomaterial is in contact with and in what state (unhydrated or hydrated) the biomaterial is introduced. These complementary reactions are described in more detail in Chapters 3 and 6–8.

The contact zone developed depends on a combination of the above-mentioned mechanisms and the actual tissue. The latter varies from a cellular-free high-content apatite tissue in the case of dental enamel via dentine to a bone structure with cellular and body liquid contact. In contact with body liquid apatite is precipitated due to the alkaline systems Ca-aluminate and Ca-silicate. The hydrogen phosphates of the body liquid—HPO_4^{2-} and $H_2PO_4^-$—are reduced to pure phosphate ions, PO_4^{3-}. Due to the extremely low solubility product of apatite ($pK_s = 10^{-58}$) precipitation of apatite occurs upon the originally precipitated nanocrystal of the main system [5].

The material can also be in contact with other implant materials such as dental crowns, dental screws, or coatings on implants. Mechanisms 1 and 2 and mechanism 6 occur at all contact areas, between the paste and other biomaterials and between the paste and tissue. Mechanisms 3–5 occur mainly at interfaces with body liquid and tissue. The CBBC material will be in contact with different tissues—enamel, dentine, and hard bone tissue and soft tissue—as well as other biomaterials' contact surfaces. The six mechanisms affect the integration differently depending on (a) what type of tissue the biomaterial is in contact with, (b) in what state (unhydrated or hydrated) the material is introduced, and (c) what type of application is aimed for (cementation, dental fillings, endodontic fillings, sealants, coatings, or augmentation products). Both pure, nanostructural, mechanically controlled integration and chemically induced integration seem plausible. This will be discussed further in Chapters 6 and 8.

When apatite is formed at the interface according to any of the reaction mechanisms 2–4 above, at the periphery of the bulk biomaterial, biological integration may start. Bone ingrowth towards the apatite allows the new bone structure to come in integrated contact with the biomaterial. The transition from tissue to biomaterial is smooth and intricate [4]. For an experimental Ca-aluminate-based system the ingrowth is shown in Fig. 2.1.

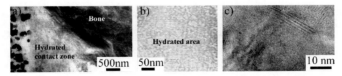

Figure 2.1   TEM image of the ceramic–bone contact zone in a sheep vertebra. Black particles are zirconia (a), STEM image of the hydrated area (b), and HRTEM image of the hydrated crystals (c). *Abbreviations*: TEM, transmission electron microscopy; STEM, scanning transmission electron microscopy; HRTEM, high-resolution transmission electron microscopy.

2.3   Conclusion and Summary

Chemically bonded ceramics have advantages related to many other biomaterials with respect to:

- the possibility of in situ–, in vivo–formed biomaterials;
- reaction temperatures close to those of tissues;
- similarity in chemistry and physics to hard tissue;
- bone and dental void–filling possibilities;
- closure of contact zones; and
- nanostructural integration.

Acknowledgement

The author thanks the personnel at Doxa Company, Sweden, and the Materials Science Department at Uppsala University for valuable inputs during a two-decade period.

References

1. Martin, R. B. (1999), Bone as a ceramic composite material, in *Bioceramics*, Ed. J. F. Shackelford (Trans Tech, Switzerland).

2. Simon, S. R., Ed. (1994), *Orthopaedic Basic Science* (Academy of Orthopaedic Surgeons, Rosemont, Amer).

3. Hermansson, L. (2011), Nanostructural chemically bonded Ca-aluminate based biomaterials, in *Biomaterials: Physics and Chemistry*, Ed. R. Pignatello (INTECH, Rijeka).

4. Hermansson, L., Lööf, J., Jarmar, T. (2009), Integration mechanisms towards hard tissue of Ca-aluminate based biomaterials, *Key Eng. Mater.*, **396–398**, pp. 183–186.

5. Axén, N., Bjurström, L-M., Engqvist, H., Ahnfelt, N.-O., Hermansson, L. (2004), Zone formation at the interface beteween Ca-aluminate cement and bone tissue environment, *Ceramics, Cells and Tissue*, 9th Annual Meeting, Faenza.

Chapter 3

Overview of Chemical Reactions, Processing, and Properties

This chapter gives an overview of chemical reactions involved in the curing of chemically bonded bioceramics (CBBCs), and how these materials are processed, and the typical property features of these biomaterials in relation to other biomaterials. A detailed description of the property profile of CBBCs will be presented in Chapters 7 and 8.

3.1 Chemical Reactions during Setting and Hardening: An Overview

Six mechanisms have been identified [1] that control how chemically bonded bioceramic (CBBC) materials are integrated onto tissue: (i) the main reaction, which includes hydration; (ii) apatite formation in the presence of phosphate ions in the biomaterial; (iii) apatite formation in the contact zone in the presence of body liquid; (iv) transformation of hydrates into apatite and other biomaterials; (v) biologically induced integration and ingrowth, that is, bone formation at the contact zone; and (vi) a mass increase reaction, especially important when unhydrated phases are used as coatings or as augmentation pastes. These mechanisms are described in next chapters.

Nanostructural Bioceramics: Advances in Chemically Bonded Ceramics
Leif Hermansson
Copyright © 2015 Pan Stanford Publishing Pte. Ltd.
ISBN 978-981-4463-43-0 (Hardcover), 978-981-4463-44-7 (eBook)
www.panstanford.com

3.1.1 Mechanism 1

Ca-aluminates (CAs) and Ca-silicates (CSs) react with water in an acid–base reaction involving dissolution of the hydrating phases into the liquid, formation of ions, and precipitation of nanosize crystals and/or amorphous phases. For CA the reaction involves the formation of katoite, $3CaOAl_2O_3 \cdot 6H_2O$, and gibbsite, $Al(OH)_3$. The overall reaction is summarised as follows:

$$3(CaO \cdot Al_2O_3) + 12H_2O \rightarrow 3Ca^{2+} + 6Al^{3+} + 4(OH)^- \rightarrow 3Ca^{2+} + 6Al(OH)_4^-$$
$$\rightarrow Ca_3[Al(OH)_4]_2(OH)_4 \text{ (katoite)} + 4Al(OH)_3 \text{ (gibbsite)}$$

The precipitation reaction is repeated until all the cement powder or the water is consumed. The crystals formed are in the nanosize range of 10–40 nm and are formed on tissue walls and in gaps and within the reacting biomaterial. This contributes to an initial mechanical integration towards tissue. Figure 3.1 shows the typical crystal size developed.

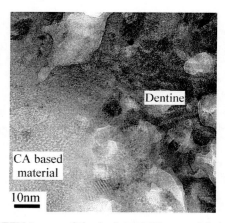

Figure 3.1   HRTEM image of the hydrated CA material showing a crystal size of approximately 25 nm and the integration to tissue (dentine). *Abbreviation*: HRTEM, high-resolution transmission electron microscopy.

3.1.2 Mechanisms 2 and 3

When phosphate ions or water-soluble phosphate compounds in the original biomaterial and/or body liquid are present, apatite formation may occur according to the following:

$$10Ca^{2+} + 3PO_4^{3-} + OH^- \rightarrow Ca_{10}(PO_4)_3OH \qquad \text{(mechanism 2)}$$

Body liquid contains hydrogen phosphate ions. In contact with the basic CA or CS materials during setting and hydration, the hydrogen phosphates are neutralised and PO_4^{3-} ions are formed:

$$HPO_4^{2-} + H_2PO_4^- + 3\,OH^- \rightarrow 2PO_4^{3-} + 3H_2O \qquad \text{(mechanism 3)}$$

Thereafter the apatite formation reaction occurs as mechanism 2:

$$10Ca^{2+} + 3PO_4^{3-} + OH^- \rightarrow Ca_{10}(PO_4)_3OH$$

Nanosize apatite crystals are precipitated on biological walls and inside the biomaterial [1]. See Fig. 3.2 below.

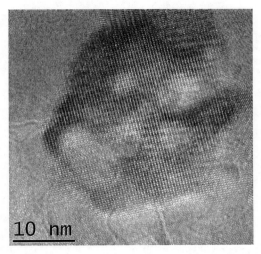

Figure 3.2   HRTEM image of a precipitated apatite crystal approximately 30 nm in size.

3.1.3   Mechanism 4

In some cases transformation of formed hydrates into apatite occurs after some time. In the main reaction in the CA system, katoite is formed as a main phase. However, in contact with body liquid, the katoite is transformed into the even more stable apatite and amorphous gibbsite. The overall reaction is as follows:

$$Ca_3 \cdot (Al(OH)_4)_2 \cdot (OH)4 \rightarrow 2Ca^{2+} + HPO_4^{2-} + 2H_2PO_4^- \rightarrow$$
$$Ca_5 \cdot (PO_4)_3 \cdot (OH) + 2Al(OH)_3 + 5H_2O$$

3.1.4 Mechanism 5

Apatite formed at the interface contributes to biological integration. Bone ingrowth towards the apatite allows the new bone structure to come in integrated contact with the biomaterial. This is an established fact for apatite interfaces [2]. For the CA system the ingrowth is shown in Fig. 3.3. Bone ingrowth towards the CA-based material allows the bone structure to come in nanostructural integrated contact with the apatite-containing biomaterial [1].

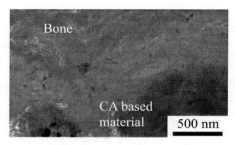

Figure 3.3   The integrated structure of CA-based material to bone tissue (rabbit, tibia).

3.1.5 Mechanism 6

When CA or CS is used as an unhydrated coating or as a paste on an implant, the body liquid uptake from the surrounding contributes to a mass increase [3], and an almost immediate 'point-welding' occurs, contributing to early improved bonding between the implant system and tissue. The water uptake can be as high as 45% according to reaction mechanism 1, written below in oxide form:

$$3CaO{\cdot}Al_2O_3 + 12H_2O \rightarrow 3CaO{\cdot}Al_2O_3{\cdot}6H_2O + 2Al_2O_3{\cdot}3H_2O$$

The actual contact zone developed depends on a combination of the above-discussed mechanisms and the tissue. The latter varies from a cellular-free high-content apatite tissue in the case of dental enamel, via dentine to a bone structure with cellular and body liquid contact. Also the material can be in contact with other implant materials such as dental crowns, dental screws, or coatings on implants. In Tables 3.1 and 3.2 in which applications and specific tissues are summarised, the demonstrated mechanisms are predominant [1].

Table 3.1 Type of tissue and possible activated mechanisms

Tissue/mechanism	No. 1	No. 2	No. 3	No. 4	No. 5	No. 6
Enamel	x	x	–	–	–	–
Dentine	x	x	x	x	(x)	
Bone	x	x	x	x	x	x

Table 3.2 Applications and possible mechanisms

Application/mechanism	No. 1	No. 2	No. 3	No. 4	No. 5	No. 6
Cementation:						
towards tissue	x	x	x	x	x	
towards implant	x	(x)				
Dental fillings:						
towards enamel	x	x				
towards dentine	x	x	x	x	(x)	
Endo fillings and bone:						
orthograde retrograde	x	x	x	x	x	
including bone	x	x	x	x	x	
augmentation products	x	x	x	x	x	(x)
Coatings:						
towards implant	x	(x)				
gap-filling	x	x	x	x	x	x

A question, related to the chemical feature often raised when CA as a biomaterial is discussed, is the issue of possible Al leakage into surrounding tissue. This has been treated in detail in Ref. [4], and from these observations the possible Al content, in addition to Al from food, etc., in blood and other organs (kidney, heart, liver, brain), is below the detection limit.

3.2 Property Features of Chemically Bonded Bioceramics

3.2.1 Property Profile Aspects

In addition to the main binding phases (cement and water), filler particles are included to contribute to some general properties of interest when used for different applications. The general contribution of added particles regards the microstructure (homogeneity aspects) and mechanical properties (especially hardness, Young's modulus, and strength). For dental applications additives are mainly glass particles. For orthopaedic application a high-density oxide is selected. For many dental applications translucency is desired, the reason why inert particles must have a refractive index close to that of the hydrates formed [5]. A preferred high-density oxide for orthopaedic application is zirconia, a material also used as a general implant material [6]. The zirconia content will contribute to the desired radio-opacity. More aspects of possible additives to CBBCs are presented in Chapter 4.

CA cements and also CS cements exhibit an inherent property not so often considered in spite of its importance for high-strength cement materials. This deals with the huge water uptake capacity of CA cements. The water consumption during hydration and curing can be as high as 45 w/o water, as compared to approximately 5 w/o for Ca-phosphates [1]. Practically this results in high-strength, low-porosity materials if an appropriate water-to-cement (w/c) ratio is selected, that is, a w/c ratio close to that of the complete reaction of the hydrating phase. This results in microstructures with essentially no large pores, just nanosize porosity between precipitated nanosize hydrates. The total residual porosity is also depending on the amounts of solid filler particles added but are often as low as 5%–10%. Figure 3.4 illustrates the general microstructure [7]. More aspects are presented elsewhere [8, 9].

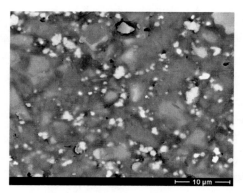

Figure 3.4   General microstructure of a CA material with a w/c ratio close to complete reaction. No porosity above 1 μm. The bright particles are the filler material, in this case ZrO_2 (white bar = 10 μm).

CA-based materials and CS-based materials exhibit significantly higher compression strength than that of Ca-phosphate-based materials or Ca-sulphate-based materials. See Table 3.3. CS-materials are foreseen to be candidate materials for trauma and for treatment of young patients, where new bone formation as a replacement of the biomaterial is aimed at [10].

Table 3.3   Typical compressive strength data of some CBC systems

Material	Ca-aluminates [11–13]	Ca-silicates [10]	Ca-phosphates [14]	Ca-sulphates [15]
Compressive strength (MPa)	150–250	100–150	<80	<50

Abbreviation: CBC, chemically bonded ceramic

Due to reduced porosity, CA materials exhibit the highest strength among CBCs. The inherent flexural strength is above 100 MPa, depending on measurement of the fracture toughness, which is about 0.7–0.8 MPa·m$^{1/2}$ and the largest distance in the microstructure of approximately 10–15 μm. The actual flexural strength is controlled by external defects introduced during handling and injection of the material. The spread in strength corresponds approximately to an *m* value of 10. The thermal and electrical properties of hydrated biomaterials are close to those of hard tissue, the reason being that the hydrates chemically belong to the same group as Ca-phosphates, the hard tissue of bone. Table 3.4 summarises results from some studies [11–14, 17, 18].

3.2.2   Practical Properties

The users of biomaterials, dentists or medical doctors, are mainly interested in a few basic properties. These properties deal with handling and safety, as well as aesthetics. The handling aspects comprise working and setting times, curing time, and consistency and interaction between the material and tools used, as well as ease of removal of excess materials after treatment. The practical properties are often not presented in the scientific literature but are found within the intellectual property and know-how of companies.

The processing aids (retarders or accelerators as well as surface-active compounds) are important for these practical properties. Examples of handling properties for biomaterials based on CA are presented in Table 3.5 [1].

Table 3.4   Mean property data of experimental dental CA-based materials

Property	Mean value
Hardness (H_v 100 g)	110–130
Young's modulus (GPa)	15–18
Compressive strength (MPa) after 28 days	180–260
Flexural strength (MPa) after 7 days	60–80
Weibull modulus, m	7–10
Thermal conductivity (W/mK)	0.8
Thermal expansion (ppm/K)	9.5
Process temperature (°C)	30–40

Table 3.5   Handling property data of CA-based biomaterials

Property	Typical value
Working time (min)	3–6
Setting time (min)	2–8
Curing time (min)	5–20*

*Can be varied within the interval depending on the selection of phases, processing agents, and the complementary binding phase

Another important property related to high-water-uptake CBBCs as CA materials is the possibility to control the dimensional change during hardening. In contrast to the shrinkage behaviour of many polymer-based biomaterials, for example, resins and poly(methyl methacrylate) (PMMA)-based biomaterials, the CAs exhibit no shrinkage but a small controlled expansion, 0.1–0.3 linear percentage [17]. This is important to avoid tensile stress in the contact zone between the biomaterial and the tissue and reduces significantly the risk of bacterial infiltration and, in the case of dental biomaterials, the chemical stability to avoid chemical attack in the oral environment. A practical consequence is reduced post-operative sensitivity. Another practical property is the radio-opacity to be able to see the biomaterial by X-ray for control after (surgical) treatment. See Fig. 3.5.

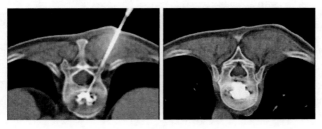

Figure 3.5   Percutaneous vertebroplasty using a CA-based biomaterial.

CS and CA materials are moisture-tolerant, meaning that no special care is needed when working in a humid environment such as the mouth.

3.3   Conclusion and Summary

CBBCs exhibit several properties suitable for in situ, in vivo placement in hard tissues. This is based on the chemical, physical, and biological features of the biomaterials.

- Six general mechanisms describe how bioceramics can be integrated onto tissue.
- CBBCs include apatite, the main chemical constituent in hard tissue.
- CBBCs tolerate moist conditions.
- The thermal and electrical properties of CBBCs are close to those of hard tissue.

Apatite is in many cases formed as an additional phase. Both pure, nanostructural mechanically controlled integration and chemically induced integration seem plausible for CBBCs when in contact with hard tissue.

Acknowledgement

The author thanks the personnel at Doxa Company, Sweden, and the Materials Science Department at Uppsala University for valuable inputs during a two-decade period.

References

1. Hermansson, L., Lööf, J., Jarmar, T. (2009), Integration mechanisms towards hard tissue of Ca-aluminate based biomaterials, *Key Eng. Mater.,* **396–398**, pp. 183–186.

2. Hench, L., and Andersson, Ö. (1993), Bioactive glasses, in *Introduction to Bioceramics*, Ed. (World Scientific, Singapore), pp. 41–62.

3. Axén, N., Engqvist, H., Lööf, J., Thomsen, P., and Hermansson, L. (2005), In vivo hydrating calcium aluminate coatings for anchoring of metallic implants, *Key Eng. Mater.,* **284–286**, pp. 831–834.

4. Axén, N., Bjurström, L-M., Engqvist, H., Ahnfelt, N.-O., and Hermansson, L. (2004), Zone formation at the interface between Ca-aluminate cement and bone tissue environment, *Ceramics, Cells and Tissue*, 9th Annual Meeting, Faenza.

5. Engqvist, H., Loof, J., Uppstrom, S., Phaneuf, M. W., Jonsson, J. C., Hermansson, L., and Ahnfelt, N.-O. (2004), Transmittance of a bioceramic calcium aluminate based dental restorative material, *J. Biomed. Mater. Res.,* **69**, pp. 94–98.

6. Adolfsson, E. (1999), *Phase Stability and Preparation of Oxide-Apatite Composites*, PhD thesis, Stockholm University (Department of Inorganic Chemistry, Sweden).

7. Hermansson, L., and Engqvist, H. (2006), Formation of nano-sized apatite coatings on chemically bonded ceramics, *Ceram. Trans.,* **172**, pp. 199–206.

8. Engqvist, H., Couillard, M., Botton, G. A., Axén, N., Ahnfelt, N.-O., and Hermansson, L. (2004), Chemical interactions between Ca-aluminate implants and bone, *28th Int. Conf. Adv. Ceram. Comp.,* 25 **4**, pp. 459–554.

9. Engqvist, H., Botton, G. A., Couillard, M., Mohammadi, S., Malmström, J., Emanuelsson, L., Hermansson, L., Phaneuf, M. W., and Thomsen, P. (2006), A new tool for high-resolution transmission electron microscopy of intact interfaces between bone and metallic implants, *J. Biomed. Mater. Res.,* **78A**, pp. 20–24.

10. Engqvist, H., Edlund, S., Gómez-Ortega, G., Loof, J., and Hermansson, L. (2006), In vitro mechanical properties of a calcium silicate based bone void filler, *Key Eng. Mater.,* **309–311**, pp. 829–832.

11. Lööf, J., Engqvist, H., Gómez-Ortega, G., Spengler, H., Ahnfelt, N.-O., and Hermansson, L. (2005), Mechanical property aspects of a biomineral based dental restorative system, *Key Eng. Mater.,* **284–286**, pp. 741–744.

12. Lööf, J., Engqvist, H., Hermansson, L., and Ahnfelt, N.-O. (2004), Mechanical testing of chemically bonded bioactive ceramic materials, *Key Eng. Mater.,* **254–256**, pp. 51–54.

13. Hermansson, L., Kraft, L., Lindqvist, K., Ahnfelt, N.-O., and Engqvist, H. (2008), Flexural strength measurement of ceramic dental restorative materials, *Key Eng. Mater.,* **361–363**, pp. 873–876.

14. Hermansson, L. (2011), Nanostructual chemically bonded Ca-aluminate based biomaterials, in *Biomaterials: Physics and Chemistry*, Ed. R. Pignatello (INTECH, Rijeka).

15. Ferrance, J. L. (1995), *Materials in Dentistry* (J.P. Lippincott Company, Philadelphia).

16. Hermansson, L., Engqvist, H. J., Gomez-Ortega, G., Abrahamsson, E., and Björklund, K. (2006), Nanosize biomaterials based on Ca-aluminate, *Key Eng. Mater.,* **49**, pp. 21–26.

17. Hermansson, L., Faris, A., Gomez-Ortega, G., Abrahamsson, E., and Lööf, J. (2009), Calcium aluminate based dental luting cement, *Ceram. Eng. Sci. Proc.,* **31**, pp. 27–38.

18. Kraft, L. (2002), *Calcium Aluminate Based Cement as Dental Restorative Materials*, PhD thesis, Uppsala University.

Chapter 4

Additives Used in Chemically Bonded Bioceramics

This chapter describes typical additives used in chemically bonded bioceramics (CBBCs). These additives comprise active complementary binders, processing agents, and fillers.

4.1 Additives Normally Used for Chemically Bonded Bioceramics

Chemically bonded bioceramics (CBBCs)—based on phosphates, silicates, or aluminates—are all depending on additives to obtain the final properties for the specific application aimed at. The additives have in general at least three important tasks in CBBCs. These deal with the main reacting binding phase, processing aspects, and the final long-term properties.

Typical additives for the main binding phase are complementary chemically bonding phases and/or glass ionomer systems, including glass ionomer glass and an acid, typically a polyacrylic acid (PAA) [1]. Additives selected to control the formation process of CBBCs include retarders or accelerators, depending on which specific main chemically bonded binding phase is used. Other process-related additives are surface-active agents to control the dispersion and homogeneity during hydration. Typical additives here are dispersing

Nanostructural Bioceramics: Advances in Chemically Bonded Ceramics
Leif Hermansson
Copyright © 2015 Pan Stanford Publishing Pte. Ltd.
ISBN 978-981-4463-43-0 (Hardcover), 978-981-4463-44-7 (eBook)
www.panstanford.com

agents and surfactants. Inert additives are introduced in the CBBCs for many different reasons but mainly to control homogeneity and improve specific properties. Typical inert additives used are stable glasses, oxides, double oxides, or hydrated phases. The use of different types of additives as well as complementary binding phases and processing aids is treated in more detail next.

4.1.1 Complementary Binding Phases for Chemically Bonded Bioceramics

The main binding phase for CBBCs and typical complementary binding phases used for the main phase are presented in Table 4.1 [1, 2].

Table 4.1 Examples of CBBCs and additional binding phases

Main system	Examples of added binding phase
Ca-silicates	Ca-aluminates, glass ionomer systems
Ca-aluminates	Other Ca-aluminates, Ca-silicates, or glass ionomer systems
Ca-phosphates	Ca-sulphates
Ca-sulphates	Ca-phosphates

The main purpose of added binding phases is to accelerate the binding process. Ca-aluminates in general react faster than Ca-silicates and Ca-phosphates. To improve early hardening glass ionomer systems can be used. The proposed CBBC systems have in general favourable biocompatible properties, so additional CBBCs do not change the established biocompatibility of the main binding phase. An attractive additive to speed up the setting and hardening of CBBC systems is glass ionomer systems. These systems contribute to early hardening due to the formation of cross-linked PAAs. Ca and Al ions cross-link the PAA, resulting in early solidification [3].

4.1.2 Processing Agents for Chemically Bonded Bioceramics

Processing agents are introduced to either retard or accelerate the binding of the main phase. Too rapid hydration will result in a temperature raise, which may damage surrounding tissue, and too slow hydration is inappropriate from a practical clinical aspect. So depending on the main system selected for a given application retarders or accelerators may be necessary to use. The processing agents also affect rheological properties and contribute to desired consistency. For Ca-aluminate-based systems a Li salt is often used as an accelerator [4]. For tri-Ca-phosphate (TCP) citric acid is often used as a retarder [2]. Retarders and accelerators often belong to the know-how of the commercial producers of biomaterials.

4.1.3 Fillers Used in Chemically Bonded Bioceramics

Important complementary additives in CBBCs are inert filler materials. The inert fillers in CBBCs are introduced to contribute to general properties related to homogeneity and microstructure and to improve mechanical properties such as strength and wear resistance. Inert fillers are also introduced to control specific properties related to the inherited property of additives. Slowly resorbable phases may influence the biologically related properties of biocompatibility, bioactivity, and antibacterial properties. In Table 4.2 below is presented the contribution of some additives to specific properties [1, 5–7].

Table 4.2   Examples of additives and specific properties achieved

Additives	Improved property of the CBBCs
Glass	Transparency
ZrO_2	Radio-opacity
Pre-hydrated phases	Transparency
Ca-phosphates	Bioactivity
SrF_2	Antibacterial properties

The amount of fillers can also be used to control the porosity of the final products [8]. CBBCs always contribute to a certain amount of nanoporosity. To reduce the total porosity inert solid fillers may be added [4].

4.2 Summary

Additives for CBBCs are introduced to promote early and controlled hardening and hydration, controlled microstructure, and homogeneity, as well as specific properties related to mechanical strength, dimensional stability, radio-opacity, transparency, and biologically related properties, including biocompatibility, bioactivity, controlled resorption, and antibacterial properties.

Acknowledgement

The author thanks the personnel at Doxa Company, Sweden, and the Materials Science Department at Uppsala University for valuable inputs during a two-decade period.

References

1. Hermansson, L. (2011), Nanostructural chemically bonded Ca-aluminate based biomaterials, in *Biomaterials: Physics and Chemistry*, Ed. R. Pignatello (INTECH, Rijeka), pp. 831–834.

2. Nilsson, M. (2003), *Injectable Calcium Sulphates and Calcium Phosphates as Bone Substitutes*, PhD thesis, Lund University.

3. Jefferies, S. R., Pameijer, C. H., Lööf, J., Appleby, D., and Boston, D. (2009), Clinical performance of a bioactive dental luting cement- a prospective clinical pilot study. *J. Clin. Dent.*, **20**, pp. 231–237.

4. Kraft, L. (2002), *Calcium Aluminate Based Cement as Dental Restorative Materials*, PhD thesis, Uppsala University.

5. Engqvist, H., Loof, J., Uppstrom, S., Phaneuf, M. W., Jonsson, J. C., Hermansson, L., and Ahnfelt, N.-O. (2004), Transmittance of a bioceramic calcium aluminate based dental restorative material, *J. Biomed. Mater. Res.*, **69**, pp. 94–98.

6. Lööf, J., Svahn, F., Jarmar, T., Engqvist, H., and Pameijer, C. H. (2008), A comparative study of the bioactivity of three materials for dental applications, *Dent. Mater.*, **24**, pp. 653–659.

7. Hermansson, L. (2012), Aspects of antibacterial properties of nanostructural calcium aluminate based biomaterials, *Proc. Adv. Ceram. Comp., Ceram. Eng. Sci. Proc.*, **33**, pp. 57–64.

8. Hermansson, L. (2010), Chemically bonded bioceramic carrier systems for drug delivery, *Ceram. Eng. Sci. Proc.*, **3**, pp. 77–88.

Chapter 5

Test Methods with Special Reference to Nanostructural Chemically Bonded Bioceramics

This chapter gives an overview of the most critical test methods used for determining important properties for nanostructural chemically bonded bioceramics (CBBCs). The test methods dealt with in the chapter comprise nanostructural evaluation, including contact zone determination and mechanical strength and dimensional stability evaluation. Test methods related to chemical stability, microleakage, biocompatibility, bioactivity, anti-bacterial properties, and others will be treated in relation to examples of nanostructures and properties in Chapters 7 and 8.

5.1 Introduction

When evaluating new biomaterials it is very important to know and understand the relationship between the new material, the specific test methods used, and what application the material is intended for. Often according to standards the methods are established and decided related to specific early commercially introduced biomaterials. A new material may be a good choice for a given application but can exhibit problems related to the test method. As an example, for a chemically bonded ceramic material based on Ca-aluminate, testing at body temperature or at least above 30°C

Nanostructural Bioceramics: Advances in Chemically Bonded Ceramics
Leif Hermansson
Copyright © 2015 Pan Stanford Publishing Pte. Ltd.
ISBN 978-981-4463-43-0 (Hardcover), 978-981-4463-44-7 (eBook)
www.panstanford.com

is critical, since this material develops different hydrated phases, depending on the test temperature [1, 2]. Often standards relate to room temperature. Another example is how mechanical strength is measured and how the test rig is designed. Size, shape, and time aspects may influence the result.

In this chapter are presented a few critical test methods used in the evaluation of new CBBCs.

5.2 Test Methods and Nanostructures

5.2.1 Micro-/Nanostructural Evaluation

Methods used in the evaluation of microstructures, and phase and elemental analyses are traditional scanning electron microscopy (SEM), transmission electron microscopy (TEM), high-resolution transmission electron microscopy (HRTEM), X-ray diffraction (XRD), X-ray photoelectron spectroscopy (XPS), and scanning transmission electron microscopy (STEM) with energy-dispersive X-ray (EDX) [3].

To analyse interfaces and calcified tissue at the highest level, TEM in combination with focused ion beam (FIB) microscopy for intact site-specific preparation of the TEM samples at very high site-specific accuracy is recommended. This procedure is treated in detail in Ref. [4]. Cross-sectional TEM samples from the interface between, for example, enamel and a dental filling material, are produced by FIB. The system scans over a beam of positively charged gallium ions over the samples, similar to an electron beam in SEM. The ions generate sputtered neutral atoms, secondary electrons, and secondary ions. Here it is possible to increase the beam current of the primary ion beam and use FIB as a fine-scale micromachining tool to cut TEM samples with high accuracy. See Fig. 5.1. To produce the TEM samples the so-called 'lift-out' technique can be used [4]. The thickness of the samples used is approximately 150 nm.

5.2.2 Mechanical Properties

Important properties for mechanical evaluation deal with compressive strength, flexure strength, Weibull modulus, Young's modulus, fracture toughness, and, for dental applications, also wear resistance [5–8].

Figure 5.1 Preparation of test samples with the FIB technique. TEM samples ready for lift-out (top), followed by mounting on a TEM grid and final polishing to electron transparency (FIB electron mode) (bottom). The samples are from the interface between bone and injectable Ca-aluminate-based material.

The materials are evaluated with regard to compressive strength (ISO 9917), biaxial flexural strength (ASTM F-394, flexural strength testing of ceramic materials), and Young's modulus (slope of compressive strength–strain curve) as a function of time when stored in water or a phosphate buffer solution at body temperature. The mechanical properties are determined after selected time periods. The typical test time is after 1 hour, 24 hour, 7 days, 4 weeks, and a selected long time period of half a year or longer.

A qualitative value of the mechanical properties is the fracture toughness, K_{IC}. This value is an intensive property and relates to the energy needed for a fracture to occur. It also relates to the possible strength of the material related to the largest defect in the stressed zone of the material. A practical consequence of the understanding of fracture toughness is therefore to achieve a consistency of the CBBC materials to avoid larger pores to be introduced during preparation and handling of the paste.

5.2.3 Dimensional Stability: Shrinkage or Expansion?

Confusion has been related to how dimensional stability should be tested for CBBCs [2, 9, 10]. This mainly relates to determination of water uptake in biomaterials and how this in general is related to expansion. For organic polymer– or glass ionomer–based biomaterials dimensional changes, especially expansion, are related to water uptake [9]. For chemically bonded bioceramics (CBBCs), where hydration is the main binding reaction, the hydration often contributes to shrinkage. However, for CBBCs, in many cases where nanostructures dominate, dimensional stability or a small expansion can occur.

Aspects related to how dimensional stability is judged are also related to test methods. This has been treated in detail in several papers [1, 9–13] and is summarised in Table 5.1.

5.3 Summary

Caution must be shown when testing CBBCs, since these low-temperature-formed biomaterials can show different behaviours at temperatures according to established standards, often room temperature, compared to those at body conditions. The temperature may affect phases formed and, furthermore, the working, setting, and curing times. Other crucial test methods relate to the geometrical stability of the in situ–, in vivo–formed CBBCs and how mechanical properties are measured.

Table 5.1 Guidelines of dimensional test evaluation for selection of correct conditions to ensure right geometry change measurement/stability

Test situation	Variation	Results	Comments
Temperature	Different temperatures	Different phases probably	Use body temperature.
Geometry	Different sizes	Depending on position and environment; the contribution from the environment higher for small test pieces	Use larger test samples.
Position	Free or restricted testing	Often higher measured expansion for free situation due to precipitation from the surrounding liquid	Use different sizes.
Environment	Water, test liquids, body liquid	Different precipitated phases probably	Surface analyses are necessary.
Test methods	Free linear expansion (laser or micrometer), optical iso-chromatic rings, split-pin method	Geometrical change, pressure obtained, indirect geometrical change	A combination of methods is often required to ensure the right understanding.

Acknowledgement

The author thanks the personnel at Doxa Company, Sweden, and the Materials Science Department at Uppsala University for valuable inputs during a two-decade period.

References

1. Kraft, L. (2002), *Calcium Aluminate Based Cement as Dental Restorative Materials*, PhD thesis, Uppsala University.
2. Midgley, H. G., and Midgley, A. (1975), The conversion of high alumina cement, *Mag. Concr. Res.*, **27**, pp. 59–77.
3. Engqvist, H., Schultz-Walz, J. E., Lööf, J., Botton, G. A., Mayer, D., Phaneuf, M. W., Ahnfelt, N.-O., and Hermansson, L. (2004), Chemical and biological integration of a mouldable bioactive ceramic material capable of forming apatite in vivo in teeth, *Biomaterials,* **25**, pp. 2781–2787.
4. Engqvist, H., Botton, G. A., Couillard, M., Mohammadi, S., Malmström, J., Emanuelsson, L., Hermansson, L., Phaneuf, M. W., and Thomsen, P. (2006), A new tool for high-resolution transmission electron microscopy of intact interfaces between bone and metallic implants, *J. Biomed. Mater. Res.,* **78A**, pp. 20–24.
5. Engqvist, H., Edlund, S., Gómez-Ortega, G., Lööf, J., and Hermansson L. (2006), In vitro mechanical properties of a calcium silicate based bone void filler, *Key Eng. Mater.,* **309–311**, pp. 829–832.
6. Lööf, J., Engqvist, H., Gómez-Ortega, G., Spengler, H., Ahnfelt, N.-O., and Hermansson, L. (2005), Mechanical property aspects of a biomineral based dental restorative system, *Key Eng. Mater.,* **284–286**, pp. 741–744.
7. Lööf, J., Engqvist, H., Hermansson, L., and Ahnfelt, N.-O. (2004), Mechanical testing of chemically bonded bioactive ceramic materials, *Key Eng. Mater.,* **254–256,** pp. 51–54.
8. Hermansson, L., Kraft, L., Lindqvist, K., Ahnfelt, N.-O., and Engqvist, H. (2008), Flexural strength measurement of ceramic dental restorative materials, *Key Eng. Mater.,* **361–363**, pp. 873–876.
9. Sunnergårdh-Grönberg, K. (2004), *Calcium Aluminate Cement as Dental Restorative*, PhD thesis, Umeå University.
10. Hermansson, L. (2011), Nanostructural chemically bonded Ca-aluminate based biomaterials, in *Biomaterials: Physics and Chemistry*, Ed. R. Pignatello (INTECH, Rijeka).

11. Kraft, L., Hermansson, L., and Gómez-Ortega, G. (2000), A method for the examination of geometrical changes in a cement paste, *RILEM Proc.,* **17**, pp. 401–413.

12. Kraft, L., and Hermansson, L. (2002), Hardness and dimensional stability of a bioceramic dental filling material, *26th Annu. Conf. Comp. Adv. Ceram.,* **23B** (American Ceramic Society).

13. Kraft, L., Engqvist, H., and Hermansson, L. (2004), Early-age deformation, drying shrinkage and thermal dilatation in a new type of dental restorative material based on calcium aluminate cement, *Cement Concrete Res.,* **34,** pp. 439–446.

Chapter 6

Why Even Difficult to Avoid Nanostructures in Chemically Bonded Bioceramics?

This chapter will present the reasons why nanostructures easily appear in many of the chemically bonded bioceramic (CBBC) systems. Complementary aspects of the chemistry, nanostructure, and properties are presented in Chapters 7 and 8.

6.1 Why Nanostructures in Chemically Bonded Bioceramics?

This question can be returned by a bold statement: It is even difficult to avoid nanostructures in chemically bonded bioceramics (CBBCs). It is very interesting that Professor Powers [1] as early as 1946 proposed ordinary Portland cement (OPC) (mainly a Ca-silicate system) to have a nanostructure. On the basis of Brunauer–Emmett–Teller (BET) (specific surface area measurement of dried, fully hydrated cement) Powers estimated the crystals to have an average diameter of approximately 14 nm. He was sorry the resolution of microscopes at that time was too low to see nanocrystals. Approximately 50 years later the nanocrystal size of hydrated Ca-silicates was confirmed by high-resolution transmission electron microscopy (HRTEM) [2].

Nanostructural Bioceramics: Advances in Chemically Bonded Ceramics
Leif Hermansson
Copyright © 2015 Pan Stanford Publishing Pte. Ltd.
ISBN 978-981-4463-43-0 (Hardcover), 978-981-4463-44-7 (eBook)
www.panstanford.com

The nanostructures developed in some of the CBBCs, especially the $CaO–Al_2O_3–H_2O$ (CAH), $CaO–SiO_2–H_2O$ (CSH), and $CaO–P_2O_5–H_2O$ (CPH) systems, are a consequence of the low-solubility product of the phases formed in these systems [3] (see Table 6.1).

Table 6.1   The solubility products of some of the phases in the CAH, CSH, and CPH systems

Phase	pK_s	System
Octacalcium phosphate	48.5	CPH
Hydroxyapatite	58.5	CPH
Fluoroapatite	59.7	CPH
Katoite	22.3	CAH
Gibbsite	32.2	CAH
Tobermorite	–	CSH

These values correspond to ion concentrations of approximately 10 mM or less. Interesting is that the solubility of apatites is the lowest of all these possible phases. This means that in the presence of phosphate ions in contact with the basic system Ca-aluminate (CA) or Ca-silicate, possible apatite formation is even likely and in fact difficult to avoid. This explains the bioactivity seen in these systems (see section 6.2 later).

Below is given a survey presentation, including calculations based on pK_s values, of how and why nanostructural phases appear. The example is from the CA system and why apatite phases appear in the contact zone between the biological surrounding and the biomaterial.

6.1.1   Calculations

The phases katoite, $Ca_3[Al(OH)_4]_2(OH)_4$, and gibbsite, $Al(OH)_3$, are formed in the CA system. For katoite with even the highest solubility of these low-solubility phases, a rough calculation shows the easiness of forming nanosize crystals.

$pK_s = 22.3$ for katoite, $Ca_3[Al(OH)_4]_2(OH)_4$

$$\rightarrow$$

Ion concentration for precipitation is approximately 10 mM.

$\rightarrow$

Approximately 6×10^{21} molecules/litre for precipitation to occur

Actual volume of interest is approximately 10^3 nm^3.

Number of molecules at saturation $6 \times 10^{21}/10^{24} \times 10^3 = 6$.

$\rightarrow$

Approximately 50–200 ions per 10^3 nm^3 = Approximately 5 ions per 10 nm

$\rightarrow$

Ions jump into the water–liquid surrounding, and immediately the condition for saturation is fulfilled and precipitation of crystals occurs.

$\rightarrow$

The surface energy of these approximately 1.5 nm molecules is extremely high, and in the hydrate systems approximately 1000 molecules in a crystal are necessary to reduce the total surface energy to stability at the environmental conditions at body temperature.

$\rightarrow$

Nanocrystals of an approximate size of 15 nm appear.

The nanostructures observed in CA and Ca-silicate systems confirm a typical size in the interval of 10–40 nm [4]. Even the tooth structure in detail reveals individual crystals of a size of approximately 20 nm [5].

In addition to phases based on the original biomaterial systems, in contact zones to organic tissue, apatite phases may be found—discussed in detail in Chapter 3. In section 6.2 next is presented a summary of possible reactions and how the stable (low-solubility phases) appear in the bulk material and in the contact zone to hard tissue. In section 6.3 is discussed the nanoporosity between the precipitated nanocrystals. The nanochanells between the nanocrystals are in the range of 1–3 nm.

6.2 Nanostructures in the Calcium Aluminate–Calcium Phosphate System

The six different mechanisms involved during hydration and curing of CBBCs are presented in Chapter 3 [6]. The biomaterials will be in contact with different tissues—enamel, dentine, and hard bone tissue and soft tissue—as well as other biomaterial contact surfaces. These six mechanisms affect the integration differently depending on (a) what type of tissue the biomaterial is in contact with, (b) in what state (unhydrated or hydrated) the CA is introduced, and (c) what type of application is aimed for (cementation, dental fillings, endodontic fillings, sealants, coatings, or augmentation products). The actual contact zone developed depends on a combination of the discussed mechanisms and the surrounding tissue. The latter varies from a cellular-free high-content apatite tissue in the case of dental enamel, via dentine to a bone structure with cellular and body liquid contact. Both pure nanostructural, mechanically controlled integration and chemically induced integration seem plausible.

Table 6.2 presents a summary of the six mechanisms involved in the integration of calcium aluminate–calcium phosphate (CAPH) materials towards tissue and implant surface.

Table 6.2 Chemical reactions of the CAPH[a] system in contact with different environments

Reaction mechanism	Description	Comments
Mechanism 1: Main reaction	$3(CaO\ Al_2O_3) + 12H_2O \rightarrow 3Ca^{2+} +$ $6Al^{3+} + 4(OH)^- \rightarrow 3Ca^{2+} + 6Al(OH)_4^-$ à $Ca_3[Al(OH)_4]_2(OH)_4$ (katoite) + $4Al(OH)_3$ (gibbsite)	Katoite and gibbsite formed as the main nanosize hydrates
Mechanism 2: Complementary reaction with phosphate-containing solution	$5Ca^{2+} + 3PO_4^{3-} + OH^- \rightarrow Ca_5(PO_4)_3OH$	Additional phase formed: nanosize apatite
Mechanism 3: Contact zone reaction with body liquid in the presence of the basic CA phase	$HPO_4^{2-} + OH^- \rightarrow PO_4^{3-} + H_2O$ Thereafter the apatite formation reaction occurs as mechanism 2 $5Ca^{2+} + 3PO_4^{3-} + OH^- \rightarrow Ca_5(PO_4)_3OH$	Nanosize apatite formation in the contact zone in the presence of body liquid

Reaction mechanism	Description	Comments
Mechanism 4: Transformation reaction of the originally formed phase katoite	$Ca_3·(Al(OH)_4)_2·(OH)_4 \rightarrow 2Ca^{2+} +$ $HPO_4^{2-} + 2H_2PO_4^- \rightarrow$ $Ca_5·(PO_4)_3·(OH) + 2Al(OH)_3 + 5H_2O$	Nanocrystals of apatite and gibbsite formed in the biomaterial towards tissue
Mechanism 5: Biologically induced integration and ingrowth	Bone ingrowth towards the apatite allows the new bone structure to come into integrated contact with the biomaterial.	New bone formation at the contact zone
Mechanism 6: Mass increase reaction due to the presence of unhydrated CA	$3CaO·Al_2O_3 + 12H_2O \rightarrow 3CaO·Al_2O_3$ $6H_2O + 2Al_2O_3·3H_2O$	Mass increase and point welding

[a]$CAPH = CaO–Al_2O_3–P_2O_5–H_2O$

Figures 6.1–6.3 below show the nanostructure of phases and porosity formed.

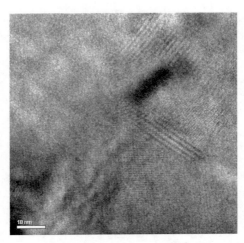

Figure 6.1   Nanostructure of CA hydrates. The pore channels are estimated to be 1–2 nm and the hydrates in the interval of 10–40 nm (white bar 10 nm).

Figure 6.1 illustrates the typical nanosize microstructure of the hydrated material with nanosize porosity between precipitated nanosize hydrates [6].

Figure 6.2 illustrates the integration between the biomaterial and the tissue, in this case dentine. Even in high magnification complete integration without gaps seems possible [7].

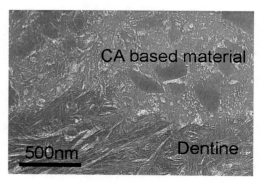

Figure 6.2   Nanostructural integration of CAPH material with dentine (grey particles in the biomaterial are glass particles).

Figure 6.3 shows how a nanocrystal of apatite is formed within the hydrated CA [6].

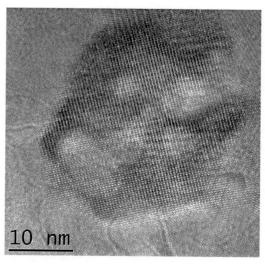

Figure 6.3   HRTEM of a precipitated hydroxyapatite crystal approximately 30 nm in size in the CAPH system.

6.3 Conclusion

CBBCs—especially materials based on phosphates, aluminates, and silicates—exhibit a general nanostructure related to both crystals and the porosity between the crystals formed. Due to a low-solubility product of the phases formed, nanocrystals are easily formed, and it is even difficult to avoid the nanostructural features. The nanocrystal size is often in the range of 10–40 nm with an open porosity with pores in the range of 1–3 nm.

Acknowledgement

The author expresses his great gratitude to all personnel at Doxa AB for input under a ten-year period.

References

1. Powers, T. C., and Brownyard, T. L. (1946), *Studies of the Physical Properties of Hardened Cement Paste*, Publication of the Research Laboratory of the Portland Cement Association, Chicago, USA.

2. Hermansson, L. (2011), Nanostructural chemically bonded Ca-aluminate based biomaterials, in *Biomaterials: Physics and Chemistry*, Ed. R. Pignatello (INTECH, Rijeka).

3. Axén, N., Bjurström, L.-M., Engqvist, H., Ahnfelt, N.-O., and Hermansson, L. (2004), Zone formation at the interface beteween Ca-aluminate cement and bone tissue environment, *Proc. Ceram., Cells Tissue, 9th Annual Meeting*, Faenza.

4. Engqvist, H., and Hermansson, L. (2006), Chemically bonded bioceramics based on Ca-aluminates and silicates, *Ceram. Trans.*, **172**, pp. 221–228.

5. Martin, R. B. (1999), Bone as a ceramic composite material, in *Bioceramics*, Ed. J. F. Shackelford (Trans Tech, Switzerland).

6. Hermansson, L., Lööf, J., and Jarmar, T. (2009), Integration mechanisms towards hard tissue of Ca-aluminate based materials, *Key Eng. Mater.*, **396–398**, pp. 183–186.

7. Engqvist, H., Schultz-Walz, J. E., Loof, J., Botton, G. A., Mayer, D., Phaneuf, M. W., Ahnfelt, N. O., and Hermansson, L. (2004), Chemical and biological integration of a mouldable bioactive ceramic material capable of forming apatite in vivo in teeth, *Biomaterials*, **25**, pp. 2781–2787.

Chapter 7

Nanostructures and Related Properties

This chapter gives an overview of nanostructures and general related properties such as contact zone closure and mechanical properties. Special attention will be paid the Ca-aluminate and Ca-silicate systems. Complementary aspects of nanostructures and specific properties are presented in Chapter 8.

7.1 Nanostructure, Including Crystal Size and Porosity Structure

7.1.1 Nanostructure, Including the Nanoporosity Developed

Ca-aluminate (CA) cements and to some extent also Ca-silicate (CS) cements exhibit an inherent property not so often considered in spite of its importance for high-strength cement materials. This deals with the huge water uptake capacity during hydration. The water consumption during hydration and curing can be as high as 45 w/o water [1]. Practically this can be utilised in development of high-strength, low-porosity materials, if an appropriate water-to-

---

Lars Kraft, Swedish Cement and Concrete Research Institute, Stockholm, Sweden (lars.kraft@cbi.se), has contributed equally to the chapter.

Nanostructural Bioceramics: Advances in Chemically Bonded Ceramics
Leif Hermansson
Copyright © 2015 Pan Stanford Publishing Pte. Ltd.
ISBN 978-981-4463-43-0 (Hardcover), 978-981-4463-44-7 (eBook)
www.panstanford.com

cement (w/c) ratio is selected, that is, a w/c ratio close to that of the complete reaction of original phases. This results in microstructures with essentially no large pores, just nanosize porosity between precipitated nanosize hydrates. The total residual porosity can be controlled to be as low as 5%–10%. See Appendix: Theoretical Model for Calculation of the Optimal Volume Share of Fillers for Reduced Total Porosity [2].

7.1.2 Microstructure and Gap (Contact Zone) Closure

The microstructure of an experimental CA-based material is shown in Fig. 7.1 The small white spots are glass particles, the light-grey areas collections of katoite ($Ca_3[Al(OH)_4]_2(OH)_4$ crystals, and the dark-grey areas are collections of gibbsite ($Al(OH)_3$) crystals [3].

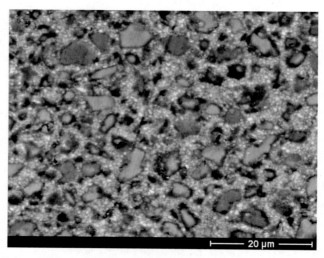

Figure 7.1   SEM micrograph of the experimental CA-based material (white bar 20 µm). *Abbreviation*: SEM, scanning electron microscopy.

The complete filling of the gaps between the material and the tooth structure is basically related to the curing mechanism of the material, which involves dissolution of the CA during the reaction with water and a precipitation of nanosize hydrates on the tooth structure. This mechanism of dissolution–precipitation is repeated during the hardening of the material. The contact zone to the tooth is filled with nanocrystals. The original gap between the restorative material and the enamel is completely filled. This contributes to mechanical nanostructural integration towards the enamel.

A magnification of the contact region between the CA-based material and enamel is shown in Fig. 7.2.

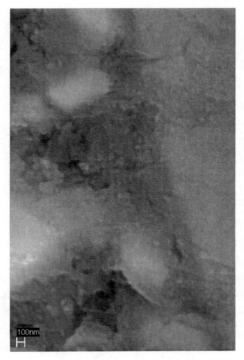

Figure 7.2   SEM micrograph of the contact zone between an experimental CA material (to the left) and the enamel (to the right); the white particles are glass particles (white bar 100 nm).

7.2   Nanostructures and Mechanical Strength

Due to reduced porosity based on the huge water uptake ability, some of the chemically bonded ceramics, CA and CS materials, exhibit high strength values [1]. The inherent flexural strength is above 100 MP based on measurement of the fracture toughness, which is about 0.7–0.8 MPa·m$^{1/2}$ [4]. The actual flexural strength is controlled by external defects introduced during handling and injection of the material.

The property data for a CA-based material—hydrated at 37°C—is presented in Table 7.1.

Table 7.1 Mechanical property data of an experimental dental CA-based material [1, 4, 5].

Property	Mean value
Hardness (H_v 100 g)	120
Young's modulus (GPa)	18
Compressive strength (MPa) after 28 days	240
Flexural strength (MPa) after 7 days	82
Weibull modulus, m	8
Fracture toughness (MPa·m$^{1/2}$)	0.75
Shear bond strength (MPa)	10

7.3 Additional Property Features of Nanostructural Chemically Bonded Bioceramics

Another important property related to nanostructural chemically bonded materials such as CA-based materials—and studied in detail in two PhD theses [5, 6]—is the possibility to avoid shrinkage and to minimise the dimensional expansion change during hardening—this in spite of the fact that formation of a single hydrate always results in shrinkage. However, for nanostructures formed in a moist environment, dimensional stability close to zero can be established, resulting in microstructures with nanocrystals and nanopores filled with water between the nanocrystals. In contrast to the shrinkage behaviour of many polymer-based biomaterials, CAs exhibit a small expansion, 0.1–0.3 linear percentage [5]. This is important to avoid tensile stress in the contact zone between the biomaterial and the tissue and reduces significantly the risk of bacterial infiltration. The nanochannels surrounding all formed nanoscrystals will also contribute to pressure relief. For the patient an extremely low post-operative sensibility has been found [7, 8]. In this context the extremely low microleakage for these chemically bonded ceramics should also be mentioned. This will be described in detail in Chapter

8. Another property related to the nanostructure is the possibility to obtain translucent materials [9].

CS materials are foreseen to be candidate materials for trauma and for treatment of young patients where new bone formation as a replacement of the biomaterial is aimed at. CS materials, as CA materials, exhibit higher compression strength that Ca-phosphate (CP) materials do, which is why load-carrying applications are foreseen for the slowly resorbable CS materials [1].

7.4 Conclusion

General properties of chemically bonded bioceramics (CBBCs) due to the nanostructures developed deal mainly with:

- high mechanical strength;
- reduced porosity;
- complete sealing of contact zones to tissue and other biomaterials due to nanostructural hydration mechanisms;
- dimensional stability; and
- possibility of load bearing biomaterial applications.

Appendix: Theoretical Model for Calculation of the Optimal Volume Share of Fillers

The model describes how to calculate the optimal vol.% of inert fillers in the reacting CBBC in order to simultaneously reduce the fraction of unreacted phases and reduce porosity [11–13].

Question: Assuming that a powder mix of cement and filler material (FM) is compacted to a certain level, how much porosity will be developed and what content of unhydrated cement will be left after the reaction?

Let

x = vol.% filler material (FM) in the powder mix

y = volume fraction of water in the compacted body

z = volume fraction of non-hydrated cement after hydration

z_0 = volume share of non-hydrated cement in the initial powder mix

$k = w/c$ (volume w/c ratio required for total hydration)

The share of unreacted cement in the compacted body, initially z_i, is under the assumption that no cement is hydrated during the compaction, as follows:

$$z_i = z_0(1 - y) \tag{7.1}$$

Since the volume share of cement that the water y can hydrate is

$$z_h = y/k \tag{7.2},$$

the fraction of unhydrated cement after hydration will be

$$z = z_i - z_h = z_i - yk \tag{7.3}$$

Taking into account that $z_0 + x = 1$, Eqs. 7.1 and 7.2 give

$$z = (1 - x)(1 - y) - y/k \tag{7.4}$$

Now the porosity in the hydrated body can be calculated as

$$\text{vol.\% pores} = y - z_h \epsilon,$$

where ϵ is a volume increase factor = cement density/hydrate density.

The hydrated cement has lower density than the unreacted cement phases, so it expands and partially fills up the pores left by the water when it reacts with the cement. Equations 7.2, 7.3, and 7.4 now yield

$$\text{vol.\% pores} = y - \epsilon[(1 - x)(1 - y) - z]$$

The k value is different for different cement types.

References

1. Hermansson, L. (2011), Nanostructural chemically bonded Ca-aluminate based biomaterials, in *Biomaterials: Physics and Chemistry*, Ed. R. Pignatello (INTECH, Rijeka).

2. Adolfsson, E. (1993), *Phase and Porosity Development in the CaO-Al$_2$O$_3$-H$_2$O System*, Materials Science thesis, Uppsala University.

3. Hermansson, L., and Engqvist, H. (2006), Formation of nano-sized apatite coatings on chemically bonded ceramics, *Ceram. Trans.*, **172**, pp. 199–206.

4. Engqvist, H., Kraft, L., Lindqvist, K., Ahnfelt, N.-O., and Hermansson, L. (2007), Flexural strength measurement of ceramic dental restorative materials, *J. Adv. Mater.*, **39**, pp. 41–45.

5. Kraft, L. (2002), *Calcium Aluminate Based Cement as Dental Restorative Materials*, PhD thesis, Uppsala University.

6. Lööf, J. (2008), *Calcium-Aluminate as Biomaterial*, PhD thesis, Uppsala University.

7. Jefferies, S. R., Appleby, D., Boston, D., Pameijer, C. H., and Lööf, J. (2009), Clinical performance of a bioactive dental luting cement- a prospective clinical pilot study, *J. Clin. Dent.*, **20**, pp. 231–237.

8. Pameijer, C. H., Zmener, O., Alvarez Serrano, S., and Garcia-Godoy, F. (2010), Sealing properties of a calcium aluminate luting agent, *Am. J. Dent.*, **23**, pp. 121–124.

9. Engqvist, H., Lööf, J., Uppström, S., Phaneuf, M. W., Jonsson, J. C., Hermansson, L., and Ahnfeldt, N.-O. (2004), Transmittance of a bioceramic calcium aluminate based dental restorative material, *J. Biomed. Mater. Res. Part B: Appl. Biomater.*, **69**(1), pp. 94–98.

Chapter 8

Nanostructures and Specific Properties

This chapter gives an overview of the importance of nanostructures, including nanocrystals and nanoporosity, in relation to specific properties achievable. The properties dealt with in this chapter comprise bioactivity, anti-bacterial properties, microleakage, haemocompatibility, and controlled drug delivery.

8.1 Nanostructures, Including Phases and Porosity for Specific Properties

Ca-aluminate cements and to some extent also Ca-silicate cements exhibit inherent properties not so often considered in spite of their importance in high-strength cement materials, anti-bacterial biomaterials, and bioactive materials. The specific interesting combination of properties for Ca-aluminate and Ca-silicate systems is the simultaneous appearing of bioactivity and bacteriostatic and anti-bacterials properties, as well as reduced microleakage. Other properties related to the nanostructure are the possibility to obtain translucent materials. Finally, nanoporosity with nanochannels below 5 nm, often in the range of 1–3 nm, can be used for controlled slow release of medicaments. The nanochannels surrounding all formed nanosize hydrates will also contribute to pressure relief. The observed haemacompatibility of the Ca-aluminate hydrated system will also be discussed.

Nanostructural Bioceramics: Advances in Chemically Bonded Ceramics
Leif Hermansson
Copyright © 2015 Pan Stanford Publishing Pte. Ltd.
ISBN 978-981-4463-43-0 (Hardcover), 978-981-4463-44-7 (eBook)
www.panstanford.com

8.1.1 Bioactivity and Anti-Bacterial Properties Simultaneously

The seemingly unlikely simultaneous appearance of bioactivity and anti-bacterial properties of Ca-aluminate- and Ca-silicate-based biomaterials is discussed below. Bioactive aspects of chemically bonded bioceramics (CBBCs) have been presented in detail in Chapters 2 and 3.

8.1.1.1 Bioactivity

In all the three main CBBC systems, Ca-phosphates, Ca-aluminates, and Ca-silicates, bioactivity is observed, and the reason for this can be summarised as follows:

- The chemical similarity between the hydrated phases in CBBCs and the apatite phase in hard tissue
- The possibility of obtaining nanostructures in CBBC systems in the same size range as that of apatite crystals in hard tissue
- The phase transformation of katoite into apatite and gibbsite
- The close contact between the biomaterial and the hard tissue due to the repeated precipitation of nanocrystals upon all surfaces, including those of hard tissue

8.1.1.2 Anti-bacterial aspects

The nanostructure, including nanoporosity, developed in some of the CBBCs systems near complete hydration conditions yields some unique properties related to how bacteriostatic and anti-bacterial properties develop in the biomaterial. Nanoporosity can also be used to control the release of drugs incorporated in the biomaterial. The background to this is that even if the total porosity is low, all porosity is open, thus allowing transport of molecules in the nanoporosity channels.

On the basis of several studies the following general reasons/ conditions have been identified which describe and to some extent explain the bacteriostatic and even anti-bacterial features of some CBBC-based biomaterials. These are summarised in Table 8.1 and will be discussed in some detail.

Table 8.1   Conditions contributing to anti-bacterial features of CBBC biomaterials

Condition	Description	Comments
pH	Acidic or alkaline pH interval	Anti-bacterial effect at pH < 6 and at pH > 9
Encapsulation	Entrapping of bacteria	Bacterial growth inhibition
Surface structure	Fastening of bacteria upon the structured surface	Bacterial growth inhibition
F ion presence	F ions act as OH ions	Anti-bacterial effect, even at neutral conditions

The surprising finding in studies recently performed [1, 2] show that the bacteriostatic and anti-bacterial properties of the Ca-aluminate biomaterial may not just be related to pH but also to the hydration procedure and the microstructure/nanostructure obtained. This also to some extent is an answer to why highly biocompatible and even bioactive biomaterials can combine apparently contradictory features such as biocompatibility, bioactivity, and apatite formation and environmental friendliness with bacteriostatic and anti-bacterial properties. The identified areas related to anti-bacterial features are discussed in more detail below.

8.1.1.2.1 *pH*

Bacteria have great problems to survive at low pH < 6 and at high pH > 9. Anti-bacterial features have been studied for CBBC Ca-aluminate-based biomaterials and Ca-silicate-based biomaterials. For pure Ca-aluminate-based materials the pH during the initial hardening is high, approximately 10.5. For Ca-silicate-based materials it is even higher, approximately 11.5. The anti-bacterial property is obvious. However, even for Ca-aluminate biomaterials combined with a glass ionomer system, where the cross-linking poly(acrylic acid) yields the system an initial low pH, approximately pH = 5 and < 7 for one hour, and then during final curing and hardening a pH of 8–9, anti-bacterial features appear. So it is quite clear that the pH condition is not the only aspect of the anti-bacterial effect in CBBC-based biomaterials.

8.1.1.2.2 *Encapsulation*

The main reactions in CBBCs involve precipitation of nanocrystals on tissue walls, in the material, and upon inert fillers and repeated precipitation until reacting phases are consumed, resulting in complete cavity-/gap-/void-filling. This reaction will guarantee that the nanostructure will be free of large pores, meaning no escape of bacteria. The nanocrystals will participate on all walls, within the liquid, and on all inert particles and on bacteria within the original volume. The formation of nanocrystals will continue till all voids are filled. The bacteria will be totally encapsulated and will be chemically inactivated. The nanosize hydrates are attached to both biological materials and other biomaterials. In Fig. 8.1 the nanostructural closure of the contact zone in a Ca-aluminate-based material to a titanium-based implant is obvious [3]. The contact zone between the biomaterial and tissue and other implant materials is kept intact as there is no shrinkage in the formation of the zone—just a slight expansion. Thus no tensile stresses develop. The reason for later opening in contact zones and bacterial invasion—reported in the dental literature for polymer composites [4]—is ascribed to the shrinkage of these biomaterials.

Figure 8.1   The nanostructure development at the contact zone between a Ti implant (top) and a Ca-aluminate hydrated paste (bottom); HRTEM (bar = 10 nm). Abbreviation: HRTEM, high-resolution transmission electron microscopy.

8.1.1.2.3 *Surface structure*

The bacteriostatic and anti-bacterial properties are in addition to pH conditions and the nanostructural entrapping mechanism also related to the surface structure developed for the hydrated biomaterial. The nanoparticle/crystal size of hydrates is in the interval of 15–40 nm, with a nanoporosity size of 1–3 nm. The number of pores per square micrometer is at least 500, preferably >1000 [1]. The number of nanopores will thus be extremely high, which will affect the possibility of catching and fastening bacteria to the hydrate surface—an analogue to how certain peptides may function as anti-bacterial material due to a structure with nanosize holes within the structure. This may also provide long-term anti-bacterial activity after the initial hydration.

8.1.1.2.4 *OH and F ions*

The size of F ions is almost the same as that of OH ions, approximately 1.4 Å. It is proposed that the anti-bacterial effect may partly be related to the presence of F ions. These may have the same effect upon the bacteria as high pH, that is, a high hydroxyl concentration. The F-containing slowly resorbable glass and the Sr-fluoride may thus contribute to the anti-bacterial features.

The nanostructure of Ca-aluminate-based materials makes these biomaterials potential delivery carriers for drugs, including antibiotics [5].

8.1.2 Microleakage

The oral environment is a body position with high bacterial activity, which causes the most common dental problem—caries formation. Marginal leakage and related possible caries are the most common reasons for replacement of a filling. More than 50% of all dental restorations today are restorations of old dental fillings [4]. This can be correlated to the contact zone between the filling material and the tooth structure, possible bonding materials, and the tooth surface structure and stresses in the contact zone.

Some of the CBBCs seem to have some general properties which should help in reducing microleakage. New chemically bonded ceramic systems have been proposed as an alternative material to those based on polymers or metals [6]. Special interest has been shown in the $CaO-Al_2O_3$ system and the $CaO-SiO_2$ system [7, 8]. Since these technologies provide materials with both small expansion and bioactivity it is hypothesised that it would result in minimal marginal leakage [9]. The main hydration reactions in the Ca-aluminate system, the general stability of the hydrates formed, and the nanostructure developed make these materials suitable as injectable biomaterials into tissue—within odontology as cements and restoratives [10, 11].

For evaluation of microleakage, restorations were ground with a straight cylinder (Horico Diamond medium). To check the materials' tendency to show leakage the fillings were subjected to 500 cycles in 5°C and 55°C water with a 'hold time' of 30 seconds. The dye penetrant (Spotcheck; Magnaflux Corp.) was then applied to the restoration surface and remained on the surface for 30 seconds before excess dye was removed. Dye penetration images of the unfilled, filled, and stressed cavities were taken using a stereo-microscope (LOM) with a digital camera connected to a frame grabber program [12]. The unstressed experimental Ca-aluminate-based filling in light microscopy is shown in Fig. 8.2.

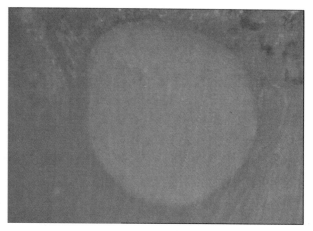

Figure 8.2   Image of an unstressed Doxa experimental filling (LOM).

None of the 16 fillings showed any dye penetration ($n = 16$) after the thermo-cycling program. The marginal leakage could be considered as zero (see Fig. 8.3).

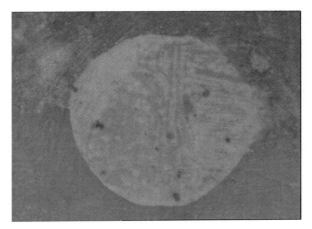

Figure 8.3   Image of stressed experimental fillings (LOM).

However, as can be seen in Fig. 8.4 the surface structure of the fillings changed somewhat. As has been earlier reported the system in pure water will exhibit a surface reaction or precipitation upon the surface. No dye penetration could be detected on the cross-sectioned fillings.

For the polymer composites, 15 out of 16 fillings showed microleakage (see Figs. 8.4 and 8.5). Most of the marginal leakage (violet colour) occurred towards the enamel side of the cavity.

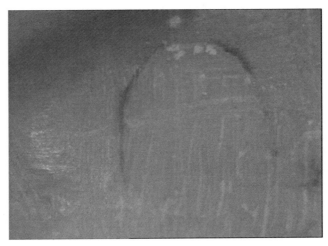

Figure 8.4   Image of stressed resin composite fillings (LOM).

Figure 8.5 Image of stressed resin composite fillings (LOM).

8.2 Drug Delivery Carriers

The open nanostructural porosity in CBBCs, including geopolymers, makes them suitable as carriers for controlled delivery of medicaments [5, 13]. This is treated in detail in Chapter 11.

8.3 Haemocompatibility

The increasing incidence of osteoporosis worldwide drives for strong development of new bone substitute materials and methods for minimally invasive fracture treatment. This trend is illustrated in the spreading clinical acceptance of vertebroplasty and kyphoplasty spinal treatments, involving stabilisation of collapsed vertebrae by percutaneous transpedicular cement injection into the damaged spinal bodies [14–16]. Acrylic bone cement—poly(methyl methacrylate) (PMMA)—is well-established and used for implant fixation, fracture treatment (including vertebrae compression fractures), and cranio-maxillo-facial applications. Also several new types of orthopaedic cements or bone void fillers which are used as pastes and which cure in vivo are commercially available. Most of these products are ceramic and based on various calcium salts such as calcium phosphates or calcium sulphates, but also new ceramic cements are entering the market [17, 18]. Calcium salt–based cements have many advantages for polymer materials; the absence of organic monomers, lower exothermal curing reactions, a higher degree of bioactivity, and for some of these materials bioresorption over time are favourable factors.

Interest in the haemocompatibility of bone cement and injectable ceramic pastes has increased because of claimed risks for adverse coagulation as the material may enter the blood system unexpectedly during injection into vertebral bodies, for example, during vertebroplasty or kyphoplasty. The clotting behaviour of some common bone substitute materials, which are used for or are candidates for vertebroplasty and kyphoplasty are presented below. The process that leads to thrombosis formation as blood contacts an artificial surface depends on a range of factors coupled to the material and its surface characteristics, the rheology, and the biological aspects commencing with the initial protein adsorption [19].

Below is presented an investigation of four orthopaedic cements: a PMMA (traditional bone cement) and three calcium-based ceramic cements, using a close-circuit Chandler loop model with the inner surfaces of the poly(vinyl chloride) (PVC) tubing coated with heparin. The model exposes the test materials to fresh human whole blood. A special procedure was developed to evaluate solidifying pastes in the Chandler loop model. This procedure covers a section of the inner wall of the tubing with a thin layer of non-cured cement paste. Thereafter the tubing is filled with fresh whole blood and the loop is closed. The loop is rotated at 32 rpm in a 37°C water bath for 60 minutes. The cements are curing in contact with the flowing blood.

After incubation, the blood and the material surfaces were investigated with special attention to clotting reactions. Blood samples were collected and supplemented with ethylene-diamin-tetra acid (EDTA) for cell count analysis. Blood from the loops was centrifuged to generate plasma for analysis of the thrombin–anti-thrombin (TAT) complex, C3a, and the terminal compact complex (TCC) complement marker. It is concluded that the clotting behaviour of the Ca-aluminate-based cement and the PMMA is considerably lower than that of the calcium phosphate and sulphate materials in these tests. More details about the test method are described in Chapter 10.

The evaluation of blood tests comprises several aspects of blood–implant interaction (see Table 8.2).

Table 8.2 Blood evaluation tests

Evaluation	Test type
Thrombosis	Occular investigation, photography, thrombosis counts
Coagulation	Visible clotting
Platelets	β-TG, platelet counts (for thrombocyte activation)
Immunology	C3a, TCC for complement activation
Fibrinogen	TAT

The materials behaved differently to exposure to flowing blood during curing. The PMMA and Ca-aluminate biocements both seemed relatively unaffected by the blood test. Both materials remained clad to the tubing inner wall. The calcium phosphate material and calcium sulphate materials, however, had interacted strongly with the blood and formed blood–ceramic material mixtures which largely filled the lumen of the tubing. The remaining material was found collected in large lumps of thrombosis, which were found loose in the tubing (see Fig. 8.6).

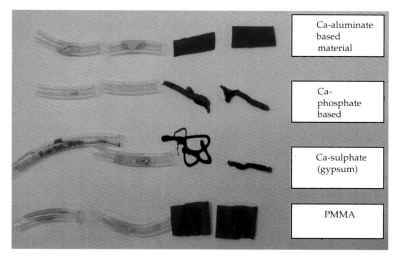

Figure 8.6 Macroscopic outcome of the blood exposure to test materials.

In summary PMMA and the Ca-aluminate biocements caused only a mild clotting behaviour, whereas the calcium sulphate and phosphate materials caused a very strong clotting reaction in these tests. A summary of the blood evaluation tests is shown in Table 8.3.

Table 8.3   Blood analysis after termination of the Chandler loop test

Material	Platelet count	C3a (μg/ml)	TCC (AU/ml)	TAT (pmol/ml)
Ca-aluminate	198–271	628–751	137–481	7.4–29.0
Ca-phosphate	8–37	–	–	–
Ca-sulphate	4–14	–	–	–
PMMA	217–308	303–444	109–126	3.4–11.6
Control	144–304	152–317	30–39	2.6–40.9
Baseline	229–335	34–417	8.5–12	– 3.6

8.4   Conclusions and Outlook

The specific interesting combination of properties for Ca-aluminate and Ca-silicate systems is the simultaneously appearing of bioactivity and bacteriostatic and anti-bacterials properties, as well as reduced microleakage. Haemocompatibility is also concluded for the Ca-aluminate hydrated system.

The anti-bacterial properties of the Ca-aluminate and Ca-silicate systems during hydration and as a permanent implant material are due to the following:

- pH during initial hardening—anti-bacterial effect at low and high pH
- F ion presence—similarity in size to that of hydroxyl ions
- The nanostructure developed—entrapping of bacteria and growth inhibition
- The surface structure of hydrated Ca-aluminate—fastening of bacteria to the surface and growth inhibition

Nanoporosity, with nanochannels below 20 nm, often in the range of 1–3 nm, can be used for controlled slow release of medicaments.

Acknowledgements

The author thanks the personnel at Doxa Company, Sweden, and the Materials Science Department at Uppsala University for valuable inputs during a two-decade period.

References

1. Hermansson, L. (2012), Aspects of antibacterial properties of nanostructural calcium aluminate based biomaterials, *Adv. Ceram. Comp., ICACC*, Daytona Beach.

2. Unosson, E., Cai, E., Jiang, J., Lööf, J., and Engqvist, H. (2012), Antibacterial properties of dental luting agents: potential to hinder the development of secondary caries, *Int. J. Dent.*, **2012**, ID 529495.

3. Axén, N., Engqvist, H., Lööf, J., Thomsen, P., and Hermansson, L. (2005), In vivo hydrating calcium aluminate coatings for anchoring of metal implants in bone, *Key Eng. Mater.*, **284–286**, pp. 831–834.

4. Mjör, I. A., Moorhead, J. E, and Dahl, J. E. (2000), *Int. Dent. J.*, **50**(6), p. 50.

5. Hermansson, L. (2010), Chemically bonded bioceramic carrier systems for drug delivery, *Ceram. Eng. Sci. Proc.*, **31**, pp. 77–88.

6. Hermansson, L. (2011), Nanostructural chemically bonded Ca-aluminate based biomaterials, in *Biomaterials: Physics and Chemistry*, Ed. R. Pignatello (INTECH, Rijeka).

7. Hermansson, L., Lööf, J., and Jarmar, T. (2009), Integration mechanisms towards hard tissue of Ca-aluminate based biomaterials, *Key Eng. Mater.*, **396–398**, pp. 183–186.

8. Engqvist, H., Edlund, H., Lööf, J., Ottosson, M., and Hermansson, L. (2006), In vitro mechanical properties of a calcium silicate based bone void filler, *Key Eng. Mater.*, **309–311**, pp. 829–832.

9. Pameijer, C. H., Zmener, O., and Serrano, S. R. (2010), Sealing properties of a calcium aluminate luting agent tested by means of a bacterial leakage test, *Am. J. Dent.*, **23**, pp. 121–124.

10. Hermanssson, L., and Kraft, L. H. (2003), Chemically bonded ceramics as biomaterials, *Key Eng. Mater.*, **247**, pp. 437–442.

11. Jarmar, T., Uhlin, T., Höglund, U., Thomsen, P., Hermansson, L., and Enqvist, H. (2008), Injectable bone cements for vertebroplasty studied in sheep vertebrae with electron microscopy, *Key Eng. Mater.*, **361–**

12. Engqvist, H., Abrahamsson, E., Lööf, J., and Hermansson, L. (2005), Microleakage of a dental restorative material based on bio-minerals, *Proc. 29th Int. Conf. Adv. Ceram. Comp.*, Cocoa Beach (*Amer. Ceram. Soc.*).

13. Forsgren, J. (2010), PhD thesis, Uppsala University (Acta Universitatis Upsaliensis).

14. Jasper, L. E., Deramond, H., Mathis, J. M., and Belkoff, S. M. (2002), Materials properties of various cements for use with vertebroplasty, *J. Mater. Sci. Mater. Med.*, **13**, pp. 1–5.

15. Kenny, S. M., and Buggy, M. (2003), Bone cements and fillers: a review, *J. Mater. Sci. Mater. Med.*, **14**, pp. 923–938.

16. Baroud, G., Bohner, M., Heini, P., and Steffen, T. (2004), Injection biomechanics of bone cements used in vertebroplasty, *Bio-Med. Mater. Eng.*, **14**, pp. 487–504.

17. Nilsson, M. (2003), *Injectable Calcium Sulphates and Calcium Phosphates as Bone Substitutes*, PhD thesis, Lund University.

18. Husband, J., Cassidy, C., Leinberry, C., and Jupiter, J. (1997), Multicenter clinical trial of Norian SRS versus conventional therapy in treatment of distal radius fractures. *Trans. Amer. Acad. Orthop. Surg.*, (Abst), p. 403.

19. Andersson, J., Sanchez, J., Ekdahl, K. N., Elgue, G., Nilsson, B., and Larsson, R. (2003), Optimal heparin surface concentration and antithrombin binding capacity as evaluated with human non-anticoagulated blood in vitro, *J. Biomed. Mater. Res. A.*, **67**(2), pp. 458–466.

Chapter 9

Dental Applications within Chemically Bonded Bioceramics

This chapter presents commercial and proposed dental applications within the biomaterials literature based on chemically bonded bioceramics (CBBCs).

9.1 Chemically Bonded Bioceramics for Dental Applications: An Introduction

The existing dental materials are mainly based on amalgam, resin composites, or glass ionomers. Amalgam, originating from the Tang dynasty in China, was introduced in the early 19th century as the first commercial dental material. It is anchored in the tooth cavity by undercuts in the bottom of the cavity to provide mechanical retention of the metal. Although it has excellent mechanical characteristics it is falling out of favour in most dental markets because of health and environmental concerns. One exception is the United States, in which amalgam still (2013) has a large market share.

The second-generation material is the resin composites, first introduced in the late 1950s. These are attached to the tooth using powerful bonding agents which glue them to the tooth structure. After technical problems over several decades, these materials today have developed to a level where they work quite well and provide excellent aesthetic results. Despite the improvements, resin composites have some drawbacks related to shrinkage, extra

Nanostructural Bioceramics: Advances in Chemically Bonded Ceramics
Leif Hermansson
Copyright © 2015 Pan Stanford Publishing Pte. Ltd.
ISBN 978-981-4463-43-0 (Hardcover), 978-981-4463-44-7 (eBook)
www.panstanford.com

bonding, irritant components, a risk of post-operative sensitivity, and technique sensitivity in that they require dry field treatment in the inherently moist oral cavity. The key problem, due to shrinkage or possible degradation of the material and the bonding, is the margin between the filler material and the tooth, which often fails over time, leading to invasion of bacteria and secondary caries. Secondary caries is a leading cause of restorative failure and one of the biggest challenges in dentistry today. As a significant number of dental restorations today are replacement of old, failed tooth fillings, it is clear that tackling this problem is a major market need [1]. Secondary caries occurs not only after filling procedures but also following other restorative procedures such as the cementation of crowns and bridges.

Glass ionomers were first introduced in 1972 and today are an established category for certain restorations and cementations. Their main weakness is the relatively low strength and low resistance to abrasion and wear. Various developments have tried to address this, and in the early 1990s resin-modified ionomers were introduced. They have significantly higher flexural and tensile strength and lower modulus of elasticity and are therefore more fracture-resistant. However, in addition to the problems of resin composites highlighted above, wear resistance and strength properties are still inferior to those of resin composites.

Alternative dental materials and implant materials based on bioceramics are found within all the classical ceramic families: traditional ceramics, special ceramics, glasses, glass-ceramics, coatings, and chemically bonded ceramics (CBCs) [2]. The CBC group, also known as inorganic cements, is based on materials in the system $CaO–Al_2O_3–P_2O_5–SiO_2$, where phosphates, aluminates, and silicates are found. Depending on in vivo chemical and biological stability, CBC biomaterials can be divided into three groups: stable, slowly resorbable, and resorbable. The choice for dental and stable materials is the Ca-aluminate (CA)- and Ca-silicate (CS)-based materials [3], discussed in detail in Chapter 3. The stable biomaterials are suitable for dental applications, long-term load-bearing implants, and osteoporosis-related applications. For trauma and treatment of younger patients, the preferred biomaterial is the slowly resorbable materials, which can be replaced by new bone tissue [4].

This chapter will discuss the relevance of using the stable chemically bonded biomaterials with regard to different dental applications.

9.2   Dental Applications

CA-based biomaterials and to some extent CSs are stable and high-strength biomaterials after hydration and can favourably be used for load-bearing applications. Ca-phosphates, Ca-sulphates, and Ca-carbonates are known to be resorbable or slowly resorbable when inserted in the body, and their main applications are within bone void-filling with low mechanical stress upon the biomaterial. The following abbreviations are normally used: C for CaO, A for Al_2O_3, S for SiO_2, P for P_2O_5, and H for H_2O.

The nature of the mechanisms utilized by CA and CS materials (especially mechanism 1, see Chapter 3), when integrating and adhering to tooth tissue and other materials, makes these materials compatible with a range of other dental materials, including resin composites, metal, porcelain, zirconia, glass ionomers, and gutta-percha. This expands the range of indications for $CaO–Al_2O_3–P_2O_5–SiO_2–H_2O$ (CAPSH) products from not only those involving tooth tissue, for example, cavity restorations, but also a range of other indications that involve both tooth tissue and other dental materials. Examples here include dental cementation, liner/base, base and core build-up, and endodontic sealer /filler materials, which involve contact with materials such as porcelain, oxides, and polymers and metals, and coatings on dental implants such as titanium- or zirconia-based materials.

9.2.1   Dental Cements

Long-term success after cementation of indirect restorations depends on retention as well as maintenance of the integrity of the marginal seal. Sealing properties of great importance deal with microleakage resistance, the retention developed between the dental cement, and the environment, compressive strength and acid resistance. Data presented below support the CA-based system to be a relevant dental cement material.

Integration with tooth tissue is a powerful feature of calcium aluminate–calcium phosphate (CAPH) system, or CA and Ca-phosphate system, and the foundation of the CAPH technology platform. Secondary caries occurs not only after filling procedures but also after other restorative procedures such as the cementation of crowns and bridges. The consequence of the difference in the mechanism of action between CAPH products and conventional products is illustrated by the study presented in detail in Ref. [5]. In Fig. 9.1 below is illustrated that the microleakage of a leading dental cement (Ketac Cem®, 3M), measured by dye penetration after thermo-cycling, was significantly higher, both before and after thermo-cycling compared to a CAPH product recently approved by the Food and Drug Administration (FDA). This has also been verified using techniques for studying actual bacterial leakage. The nanostructural precipitation upon tissue walls, biomaterials, and the original CA paste is the main reason for this, as well as the high acid corrosion resistance [6].

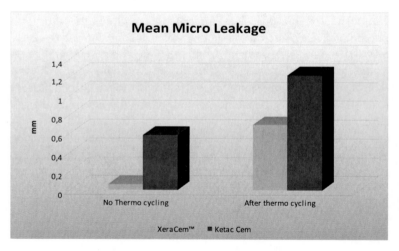

Figure 9.1   Microleakage of a CA-based cement (XeraCem, now Ceramir® C&B) and Ketac Cem.

General properties of the CAPH system used as a dental cement [7, 8] are summarised in Table 9.1.

Table 9.1   Selected properties and test methods according to SO 9917-1

Material	Net setting time (min)	Film thickness (μm)	Compressive strength (MPa)	Crown retention (kg/force)
Xeracem, Ceramir® C&B	4.8	15	196 (at 30 days)	38.6
Ketac Cem	–	19	–	26.6
RelyX-Unicem	–	–	157	39.4

A feasibility study of the CAPH material was performed in 2007. After one year all restorations were intact. A clinical study comprising 35 cemented crowns at Kornberg School of Dentistry, Temple University, shows after a three-year follow-up excellent data and feedback from participating dentists [9]. Results in details are presented in Chapter 11.

9.2.2   Endodontics

Already in the 1970s, CA was suggested as a biomaterial and tested in vivo. Hentrich et al. [10] compared CA with alumina and zirconia in an evaluation of how the different ceramics influenced the rate of new bone formation in femurs of rhesus monkeys. Hamner et al. [11] presented a study in which 22 CA roots were implanted into fresh natural tooth extraction sites in 10 baboons for periods ranging from 2 weeks to 10 months. In both studies CA successfully met the criteria for tissue adherence and host acceptance.

In a review of the biocompatibility of dental materials used in contemporary endodontic therapy [12], amalgam was compared with gutta-percha, zinc oxide-eugenol (ZOE), polymers, glass ionomer cements (GICs), composite resins, and mineral trioxide aggregate (MTA), CS-based material. A review [13] of clinical trials of in vivo retrograde obturation materials summarised the findings. GICs appeared to have the same clinical success as amalgam, and orthograde filling with gutta-percha and sealer was more effective than amalgam retrograde filling. Retrograde fillings with composite and Gluma, EBA cement, or gold leaf were more effective than amalgam retrograde fillings. However, none of the clinical trials reviewed in Ref. [13] included MTA (a CS-based biomaterial). In a 12-week microleakage study, the MTA performance was questioned compared to that of both amalgam and a composite [14].

The CAPH-based materials belong to the same material group as MTA, the CBCs [15]. MTA is a CS-based cement having bismuth oxide as the filler material for improved radio-opacity, whereas the CA material consists of the phases CA and CA_2 with zirconia as the filler material. MTA is claimed to prevent microleakage, to be biocompatible, to regenerate original tissues when placed in contact with the dental pulp or periradicular tissues, and to be anti-bacterial. The product profile of MTA describes the material as a water-based product, which makes moisture contamination a non-issue [16]. CA-cement-based materials are more acid-resistant than CS-based materials and in general show higher mechanical strength than CS materials. A five-year retrospective clinical study of CA-based material has been conducted [3]. The study involved patients with diagnosis of either chronic perapical osteitis (cpo), chronic perapical destruction, or trauma. Surgery microscope was used in all cases. For orthograde therapy the material was mixed with a solvent into an appropriate consistency and put into a syringe, injected, and condensed with coarse gutta-percha points. Machine burs were employed for root canal resection. For the retrograde root fillings (rf's), the conventional surgery procedure was performed. The apex was detected with a surgery microscope and rinsed and prepared with an ultrasonic device. Crushed water-filled CA tablets were then inserted and condensed with dental instruments. The patients' teeth were examined with X-ray, and three questions regarding subjective symptoms were put to patients: 1. Have you had any persistent symptoms? 2. Do you know which tooth was treated? 3. Can you feel any symptoms at the tooth apex?

In 13 of the 17 treated patients the diagnosis was cpo. These were treated with retrograde rf therapy. Three patients suffered from trauma or chronic perapical destruction, and these patients were treated with orthograde therapy. Out of 17 patients (22 teeth) treated, 16 patients (21 teeth) were examined with follow-up X-ray after treatment and also after two years or more. The additional patient was asked about symptoms. The results of both the clinical examination and the subjective symptoms were graded into different groups related to the success of the therapy. The results are shown in Table 9.2. A detailed presentation of the clinical study is shown in Chapter 12.

Table 9.2   Summary of the results (scores 1 and 2 considered successful; scores 3 and 4 considered failure)

	1 **Complete healing**	**2** **Incomplete healing**	**3** **Uncertain**	**4** **Failure**
No. of teeth	18	3	–	1
Percentage	82	14	–	4

Figures 9.2 and 9.3 show examples of the X-ray examination of orthograde and retrograde treatments.

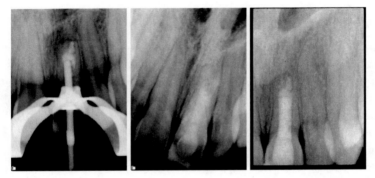

Figure 9.2   Tooth 21. Condensing with a gutta-percha pointer (left), just after treatment (middle), and at two years' control (right).

In summary, 21 out of 22 treated teeth have acceptable results, being either symptom-free or judged healed after clinical examination. The single failure can probably not be attributed to the material but rather to the difficulty of treating and sealing a multichannelled tooth. The use of CAs as root canal sealers is indirectly supported in *Introduction to Dental Materials* by van Noort [17], where the following material characteristics are looked for: biocompatible, dimensionally stable, anti-bacterial, and bioactive. As presented earlier (Chapters 3, 7, and 8) the property profile of CA-based biomaterials meets all these characteristics.

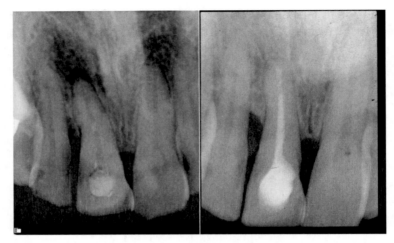

Figure 9.3 Tooth 21 at treatment (left) and at two years' control (right).

9.2.3 Dental Fillings

An important feature of the hydration mechanisms of CA-based materials is the nanostructural integration with and the high shear strength developed towards dental tissue. This makes both undercut (retention) techniques and bonding techniques redundant. The CAPH approach to dental-filling techniques is new. With the CAPH technique, the chemical reactions cause integration when the bioceramic material is placed in the oral cavity at body temperature and in a moist treatment field. Figure 9.4 shows a transmission electron microscopy (TEM) illustration of the interface between the CA-based material and dentine. This establishes a durable seal between the bioceramic and the tooth. Whereas amalgam attaches to the tooth by mechanical retention and resin-based materials attach by adhesion, using bonding agents, etchants, light-curing techniques, or other complementary techniques, the CA materials integrate with the tooth without any of these, delivering a quicker, simpler, and more robust solution.

The general aspects of CA-based materials have been presented in two PhD theses publications. Important aspects of CA materials as dental-filling materials are dealt with, such as dimensional stability, acid corrosion and wear resistance, and biocompatibility and mechanical properties [18, 19].

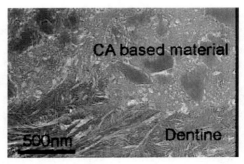

Figure 9.4   Nanostructural integration of CAPH material with dentine (grey particles in the biomaterial are glass particles).

9.2.4 Dental Implant Coatings

For successful implantation of implants in bone tissue, early stabilisation is of great importance [20]. Even small gaps may lead to relative micromotions between the implant and the tissue, which increases the risk of implant loosening over time due to formation of zones of fibrous tissues at the implant–tissue interface. Early loading of implants is of particular interest for dental implants [21]. The use of surface coatings technology is today an established method to reduce the problem with poor interfacial stability for implants. With coating technology, structural characteristics of the implant (e.g., strength, ductility, low weight, or machinability) may be combined with surface properties promoting tissue integration [22]. There are several established coating deposition techniques, for example, physical vapour deposition (sputtering) and thermal spraying techniques [23, 24]. Coatings based on calcium phosphates are the most used ones.

The implant-coating technique is used in similar ways for dental and orphopaedic applications. Further aspects of coating of implants and testing are presented in Chapter 10.

9.3   Conclusion and Summary

Chemically bonded bioceramics (CBBCs), also known as inorganic cements, are based on materials of the CAPSH system comprising phosphates, aluminates, and silicates.

It is concluded that a potential use of CAPSH materials for dental applications is based on the following features: nanostructural integration with tissue, possible apatite formation, and a mass increase yielding early point welding between the biomaterial and surrounding tissue. Consequences of nanostructural contact integration of the CAPSH system are reduced risk of secondary caries and restoration failure and reduced post-operative sensitivity. These powerful features and benefits are summarised in Table 9.3.

Table 9.3 Features and benefits of the CAPH technology platform for dental applications

• Nanostructural integration and apatite formation	• Reduced risk of secondary caries
• No shrinkage	• No or limited post-operative sensitivity
• Integration/stability/strength	• Longevity/durability
• No bonding/dry field required	• Easy and fast
• Variable consistency and compatibility to other materials	• Broad spectrum of usage with products targeting indication needs
• Moisture-tolerant	• Environmental friendliness

The following product areas have been identified on the basis of experimental material data, pre-clinical studies, pilot studies, and ongoing clinical studies: dental cements, endodontic products (orthograde and retrograde), sealants, restoratives, underfillings, and pastes for augmentation and dental implant coatings.

Acknowledgement

The author thanks the personnel at Doxa Company, Sweden, and the Materials Science Department at Uppsala University for valuable inputs during a two-decade period.

References

1. Mjör, I. A., Moorhead, J. E, and Dahl, J. E. (2000), *Int. Den. J.*, **50**(6), p. 50.

2. Ravagliolo, A., and Krajewski, A. (1992), *Bioceramics* (Chapman and Hall, London).

3. Hermansson, L. (2011), Nanostructural chemically bonded Ca-aluminate based biomaterials, in *Biomaterials: Physics and Chemistry*, Ed. R. Pignatello (INTECH, Rijeka).

4. Nilsson, M. (2003), *Injectable Calcium Sulphates and Calcium Phosphates as Bone Substitutes*, PhD thesis, Lund University.

5. Pameijer, C. H., Jefferies, S., Lööf, J., and Hermansson, L. (2008), Microleakage evaluation of Xeracem in cemented crowns, *J. Dent. Res.,* **87**(B), p. 3098.

6. Engqvist, H., Abrahamsson, E., Lööf, J., and Hermansson, L. (2005), Microleakage of a dental restorative material based on bio-minerals, *Proc. 29th Int. Conf. Adv. Ceram. Comp.*, Cocoa Beach (*Amer. Ceram. Soc.*).

7. Pameijer, C. H., Zmener, O., Garcia-Goday, F., and Alvarez-Serrano, J. S. (2009), Sealing of Xeracem and controls using a bacterial leakage model, *J Dent. Res.,* **88**(A), p. 3145.

8. Pameijer, C. H., Jefferies, S., Lööf, J., and Hermansson, L. (2008), A comparative crown retention test using XeraCem, *J Dent. Res.,* **87**(B), p. 3099.

9. Jefferies, S. R., Pameijer, C. H., Appleby, D. C., Boston, D., and Lööf, J. (2013), A bioactive dental luting cement - Its retentive properties and 3-year clinical findings. *Compend. Contin. Educ. Dent.,* **34**(Spec No 1.), pp. 2–9.

10. Heintricht, R. L., Graves, G. A., Stein, H. G., and Bajpai, P. K. (1971), An evaluation of inert and resorbable ceramics for future clinical applications, *J. Biom. Res.,* **5**(1), 25–51.

11. Hammer, J. E., Reed, M. R., and Gruelich, R. C. (1972), Ceramic root implantation in baboons, *J. Biom. Res.,* **6**(4), 1–13.

12. Haumann, C. H. J., and Love, R. M. (2003), Biocompatibility of dental materials used in contemporary endodontic therapy: a review. Part 2. Root–canal–filling materials. *Int. Endo. J.,* **36**, pp. 147–160 (Chapman and Hall).

13. Niederman, R., and Theodosopoulou, J. N. (2003), Review: a systematic review of in-vivo retrograde obturation materials. *Int. Endo. J.,* **2**(36), pp. 577–585, 1419–1423.

14. Alamo, H. L., Buruiana, R., Schertzer, L., and Boylan, R. J. (1999), A comparison of MTA, Super-EBA, composite and amalgam as root-end filling materials using a bacterial microleakage model, *Int. Endod. J.,* **32**, pp. 197–203.

15. Hermanssson, L., and Kraft, H. (2003), Chemically bonded ceramics as biomaterials, *Key Eng. Mater.*, **247**, pp. 437–442.

16. Dentsply Information (2003), *ProRoot MTA* White Brochure B18b.

17. Noort, R. V. (1994), *Introduction to Dental Materials* (Mosby, London).

18. Kraft, L. (2002), *Calcium Aluminate Based Cement as Dental Restorative Materials*, PhD thesis, Uppsala University.

19. Lööf, J. (2008), *Calcium-Aluminate as Biomaterial: Synthesis, Design and Evaluation*, PhD thesis, Uppsala University.

20. Ellingsen, J. E., and Lyngstadaas, S. P., Eds. (2003), *Bioimplant Interface, Improving Biomaterials and Tissue Reactions* (CRC Press, Taylor & Francis, New York).

21. Vercaigne, S., Wolke, J. G., Naert, I., and Jansen, J. A. (1998), Bone healing capacity of titanium plasma-sprayed and hydroxylapatite coated oral implants, *Clin. Oral Implants Res.*, **9**, p. 261.

22. Liu, X., Ding, C., Wang, Z. (2001), Apatite formed on the surface of plasma sprayed Wollastonite coating immersed in simulated body fluid, *Biomaterials*, **22**, pp. 2007–2012.

23. Rickerby, D. S., and Matthews, A. (1991), *Advanced Surface Coatings* (Chapman and Hall, New York).

24. Schneider, J. M., Rohde, S., Sproul, W. D., Matthews, A. (2000), Recent developments in plasma assisted physical vapour deposition, *J. Phys. D: Appl. Phys.*, **33**, pp. 173–186.

Chapter 10

Orthopaedic Applications within Nanostructural Chemically Bonded Bioceramics

This chapter presents commercial and proposed orthopaedic applications within the biomaterials literature based on nanostructural chemically bonded bioceramics.

10.1 Biomaterials for Orthopaedic Applications

Orthopaedic surgery comprises many different diseases and shortcomings related to bone structures. For fragility fractures, biomaterials are still sparsely used for bone repair. Autografting (bone from patients themselves) is the preferred treatment. Drawbacks related to autografts are, among others, the need for a second surgical exposure to collect tissue and a limited amount of autografts available. To solve the problems of supply, synthetic bone is one way. Especially interesting is the use of minimally invasive techniques with the possibility of in situ hardening, filling of any bone void geometry, and reduced operation time and reduced risk of complications related to infection and pain. Within biomaterials research and tissue engineering chemically bonded bioceramics (CBBCs) are of great interest.

Nanostructural Bioceramics: Advances in Chemically Bonded Ceramics
Leif Hermansson
Copyright © 2015 Pan Stanford Publishing Pte. Ltd.
ISBN 978-981-4463-43-0 (Hardcover), 978-981-4463-44-7 (eBook)
www.panstanford.com

10.2 Chemically Bonded Bioceramics for Orthopaedic Applications

As discussed in earlier chapters CBBCs are found within several material groups: sulphates, phosphates, aluminates, and silicates. As early as 1892 the first information of a Ca-sulphate (plaster of Paris, gypsum) for treatment of cavities in bone was reported. The advantages with Ca-sulphate are excellent biocompatibility and a dissolution rate suitable for certain drug release but with disadvantages such as too rapid resorption and low mechanical strength. In 1832 Ostermann prepared a Ca-phosphate paste to set in situ, but not until 1986 did the modern synthetic work on Ca-phosphate technology start by mixing an acid (water, orthophosphatic acid, or tetracalcium phosphate) with a base (dicalcium phosphate anhydrate or dehydrate phosphate). Hydroxyaptite is precipitated [1]. The apatite-based biomaterials are excellent for bone void-filling but still lack the strength for load-bearing applications. Other aspects of the use of apatite or Ca-phosphate biomaterials are related to handling, hardening time, and shelf life. This is studied, among others thing, in a PhD thesis by Åberg [2]. Biomaterials based on Ca-sulphate and Ca-phosphate are described in a PhD thesis by Nilsson [3]. Worth mentioning is the early work on Mg-aluminate-based materials by Prof. McGee [4]. However, many of these biomaterials are not typical nanostructural materials, and in some cases just one of the phases in the multiphase mateial is nanostructural.

For load-bearing applications within orthopaedics, nanostructural Ca-aluminates and Ca-silicates with higher strength than other CBBCs may be options [5–9]. Below are presented results from a relatively new CBBC material, the Ca-aluminate-based biocement.

10.2.1 Ca-Aluminate-Based Orthopaedic Materials

Within orthopaedics the following areas for Ca-aluminate biomaterials have been identified: percutaneous vertebroplasty (PVP) and kypho-vertebroplasty (KVP) and general augmentation [10].

The rate of the hydration is controlled by (1) the cement phase, (2) the particle size of the cement, (3) the hydration temperature, and (4) processing agents, including, especially, accelerators. The high water consumption feature is important for the strength but also for handling properties. The hydration rate in the Ca-aluminate system needs to be controlled to avoid a too high temperature raise. This is especially important when larger amounts of injectable materials are necessary, as for some orthopaedic applications. The PVP and KVP techniques are shortly described below.

10.2.1.1   PVP

PVP is predominantly performed under local anaesthesia by interventional radiologists. The material is injected directly into the fractured vertebra to stabilise it and relieve pain.

10.2.1.2   KVP

KVP is predominantly performed under anaesthesia by orthopaedic surgeons and neurosurgeons but increasingly also by interventional radiologists. The fracture of the collapsed vertebra is reduced by inflating a balloon (or inflatable balloon tamp [IBT]) inside it. After deflation and removal of the balloon the stabilising material is injected.

Current bone cements are based on poly(methyl methacrylate) (PMMA). Although PMMA is recognised as a successful filler material, there are shared concerns related to the following identified aspects [11, 12]:

- No biological potential to remodel or integrate with surrounding bone
- No direct bone apposition
- High polymerisation temperature
- Potential monomer toxicity
- Low radio-opacity

The Ca-aluminate-based material is highly mouldable, applicable to orthopaedic cavities with standard syringes and needles [5]. The paste cures within about five minutes at 37°C and develops strength values comparable to PMMA bone cement. The cured material is stable in the long term and shows promisingly good biocompatibility. The benefits of injectable ceramic biomaterials based on Ca-aluminate related to orthopaedic applications are as follows:

During the surgical procedure [13, 14]

- High radio-opacity allows for superior visibility of the cement and increases the probability to detect potential leakages during injection (see Fig. 10.1 below).
- High and linearly increasing viscosity reduces the risk of leakage and gives predictable handling.
- High cohesiveness optimises the cement's filling pattern in the vertebrae.
- There are no toxic or smelling fumes.

After the surgical procedure [15, 16]

- Mechanical strength
- Biocompatibility, including integration
- Long-term stability, that is, non-resorbable systems

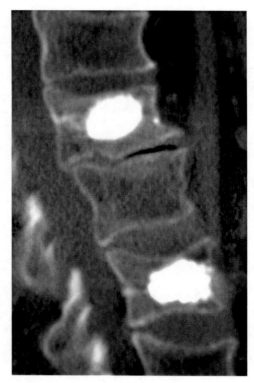

Figure 10.1 Vertebral compression factures, restored by Ca-aluminate-based material.

The injectability of chemically bonded ceramics is mainly controlled by the added water as the reacting phase with the powdered cements. This reaction is an acid–base reaction where water acts as a weak acid and the cement powder as a base. The technological importance of this is that all the water needed for injectability can be consumed in the formation of solid phases, yielding products with low porosity, one of the requirements for high strength—details in Chapter 3.

10.2.2 Ca-Aluminate-Based Orthopaedic Coating Materials

For successful implantation of implants in bone tissue, early stabilisation is of great importance [17]. Even small gaps may lead to relative micromotions between the implant and the tissue, which increases the risk of implant loosening over time due to the formation of zones of fibrous tissues at the implant–tissue interface. Early loading of implants is of particular interest for dental implants [18]. The use of surface-coating technology is today an established method to reduce the problem of poor interfacial stability for implants. With coating technology, structural characteristics of the implant (e.g., strength, ductility, low weight, or machinability) may be combined with surface properties promoting tissue integration [19]. There are several established coating deposition techniques, for example, physical vapour deposition (PVD) (sputtering) and thermal spraying techniques [20, 21]. Among CBBCs, Ca-phosphates are the most used coating materials.

This section deals with coatings deposited with established methods, with the aim of improving particularly the early-stage anchoring of metal implants to bone tissue by exploring in vivo hydration of coatings or pastes based on chemically curing ceramics. The study focuses on Ca-aluminate in the form of coatings and paste. Results are presented from an implantation study with flame-sprayed coating on titanium implants and uncoated implants augmented with a Ca-aluminate paste in the hind legs of rabbits. Implants were applied with the paste composed of a mixture of $CaO \cdot Al_2O_3$ and $CaO \cdot 2Al_2O_3$. The paste was applied manually as a thin layer on the threaded part of the implant just before implantation. The uncoated and coated implants were sterilised with hot dry air at 180°C for two hours. Female albino adult New Zealand white rabbits with a body weight of around 2.5 kg were used. Each animal received four implants, two in each hind leg. Implants were placed in the distal femoral metaphysis as well as in the proximal tibial metaphysis. Surgery followed standard procedure. The implants were screwed into predrilled and threaded cavities. Necropsy took place after 24 hrs, 2 weeks, and 6 weeks [22]. No negative effects of the implants on the general welfare of the animals were observed. The healing progressed in a normal and favourable way. As for the removal torque recordings, all Ca-aluminate coating types provided an improved implant anchoring to bone tissue after in vivo hydration as compared to that of pure metal implants. Implants on the tibia and femur sides of the knee gave similar removal torques. Table 10.1 provides average values from both tibia and femur sides.

Table 10.1   Removal torque (Ncm) for Ca-aluminate-based implants in rabbit hind legs (tibia and femur)

Implant type	24 hours	(*n*)	2 weeks	(*n*)	6 weeks	(*n*)
Flame-spraying	7.0	(8)	7.0	(8)	25	(6)
Paste augmentation	6.6	(8)	15	(6)	13	(4)
RF-PVD	12	(4)	–	–	10	(4)
Uncoated reference	3.8	(8)	5.7	(6)	14	(4)

Abbreviation: RF, radio frequency.

Twenty-four hours after implantation, Ca-aluminate between the implant and the tissue increased the removal torque to about double that of the uncoated reference implants, independent of the means of application (coatings or paste). This is considered to be attributable to the point-welding according to integration mechanism 6 described in detail in Chapter 3. Two weeks after implantation, implants combined with paste augmentation provide the highest removal torque; flame-sprayed coatings also improve the torque relative to the uncoated system. Six weeks after implantation, all systems are relatively similar (considering the uncertainty due to scatter and statistics), apart from the sprayed system, which shows significantly higher values.

10.2.2.1 Point-welding

An important aspect related to in vivo coating techniques is the possibility with 'point-welding' [22, 23]. This is defined as the possibility of early adaption to the implant surface and the hard tissue by increase of the coating material volume by interaction with the contact liquid. The coating material has a certain amount of unreacted CBBC, and this part includes body liquid into the formed contact zone material. This contributes to an early fill-up of the void between the original coating material and the tissue wall. This is illustrated in Figs. 10.2–10.4 below.

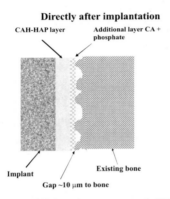

Figure 10.2   Point-welding using unreacted CBBC material—directly after implantation.

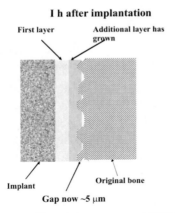

Figure 10.3   Point-welding using unreacted CBBC material—one hour after implantation.

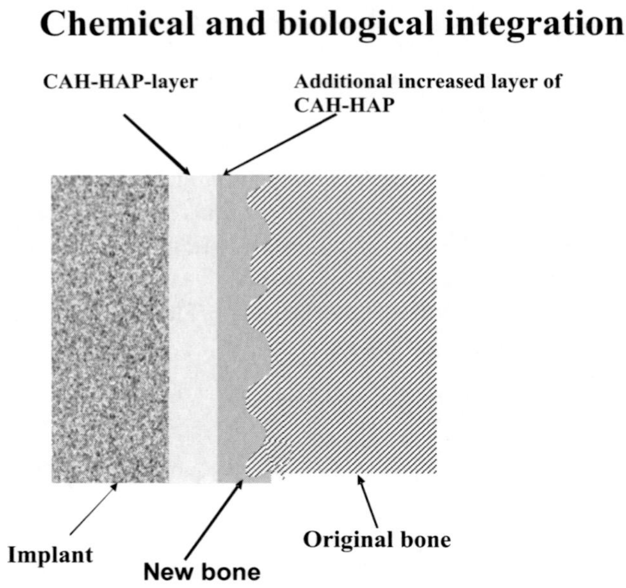

Chemical and biological integration

Figure 10.4 Point-welding using unreacted CBBC material—chemical and biological integration completed.

10.3 Summary and Conclusions

Injectable nanostructural CBBCs are still in only limited use for orthopaedic applications. Focus has been on some Ca-phosphate and Ca-sulphate systems, and experimental development activities are found based on Ca-silicates and Ca-aluminates.

The benefits of injectable ceramic biomaterials based on Ca-aluminate related to orthopaedic applications are:

- High radio-opacity allowing for superior visibility of the cement and increasing the probability to detect potential leakages during injection
- High and linearly increasing viscosity reducing the risk of leakage and giving predictable handling
- High cohesiveness optimising the cement's filling pattern in the vertebrae
- No toxic or smelling fumes
- Mechanical strength

- Biocompatibility, including integration
- Long-term stability, that is, non-resorbable systems

References

1. Bohner, M. (2000), Calciumorthophosphates in medicine: from ceramics to calcium phosphate cements, *Injury,* **31**(Suppl 4), pp. 37–47.

2. Åberg, J. (2011), *Premixed Acidic Calcium Phosphate Cements,* PhD thesis, Uppsala University.

3. Nilsson, M. (2003), *Injectable Calcium Sulphates and Calcium Phosphates as Bone Substitutes,* PhD thesis, Lund University.

4. Roemschildt, M. L., McGee, T., and Wagner, S. D. (2003), Novel calcium phosphate compositions for bone cement, *J. Mater. Sci.,* **14**, pp. 137–141.

5. Engqvist, H., Edlund, S., Gómez-Ortega, G., Loof, J., and Hermansson, L. (2006), In vitro mechanical properties of a calcium silicate based bone void filler, *Key Eng. Mater.,* **309–311**, pp. 829–832.

6. Faris, A., Engqvist, H., Lööf, J., Ottosson, M., and Hermansson, L. (2006), In vitro bioactivity of injectable ceramic orthopaedic cements, *Key Eng. Mater.,* **309–311**, pp. 833–836.

7. Engqvist, H., Persson, T., Lööf, J., Faris, A., and Hermansson, L. (2008), Chemical stability of a novel bioceramic for stabilisation of vertebtal compression, *Trends Biomater. Artif. Organs,* **21**, pp. 98–106.

8. Lööf, J., Faris, A., Hermansson, L., and Engqvist, H. (2008), In vitro biomechanical testing of two injectable materials for vertebroplasty in different synthetic bone, *Key Eng. Mater.,* **361–363**, pp. 369–372.

9. Breding, K., and Engqvist, H. (2008), Strength and chemical stability due to aging of two bone void filler materials, *Key Eng. Mater.,* **361–363**, pp. 315–318.

10. Hermansson, L. (2011), Nanostructural chemically bonded Ca-aluminate based biomaterials, in *Biomaterials: Physics and Chemistry,* Ed. R. Pignatello (INTECH, Rijeka).

11. Liebermann, I. H., Togawa, D., and Kayanja, M. M. (2005), Vertebroplasty and kyphoplasty: filler materials, *Spine J.,* **5**, pp. 305–316.

12. Heini, P. F., and Berlemann, U. (2001), Bone substitutes in vertebroplasty, *Eur. Spine J.,* **10**, pp. 205–213.

13. Engqvist, H., Couillard, M., Botton, G. A., Phaneuf, M. P., Axén, N., Ahnfelt, N.-O., and Hermansson, L. (2005), In vivo bioactivity of a novel mineral

based orthopaedic biocement, *Trends Biomater. Artif. Organs*, **19**, pp. 27–32.

14. Lööf, J. (2008), *Calcium-Aluminate as Biomaterial: Synthesis, Design and Evaluation*, PhD thesis, Uppsala University (Faculty of Science and Technology, Sweden).

15. Jarmar, T., Uhlin, T., Höglund, U., Thomsen, P., Hermansson, L., and Engqvist, H. (2008), Injectable bone cements for Vertebroplasty studied in sheep vertebrae with electron microscopy, *Key Eng. Mater.*, **361–363**, pp. 373–376.

16. Engqvist, H., and Hermansson, L. (2006), Chemically bonded nano-size bioceramics based on Ca-aluminates and silicates, *Ceram. Trans.*, **172**, pp. 221–228.

17. Ellingsen, J. E., Lyngstadaas, S. P., Eds., (2003), *Bioimplant Interface, Improving Biomaterials and Tissue Reactions* (CRC Press, Taylor & Francis , New York).

18. Vercaigne, S., Wolke, J. G., Naert, I., and Jansen, J. A. (1998), Bone healing capacity of titanium plasma-sprayed and hydroxylapatite coated oral implants, *Clin. Oral Implants Res.*, **9**, pp. 261.

19. Liu, X., Ding, C., and Wang, Z. (2001), Apatite formed on the surface of plasma sprayed Wollastonite coating immersed in simulated body fluid, *Biomaterials*, **22**, pp. 2007–2012.

20. Rickerby, D. S., and Matthews, A. (1991), *Advanced Surface Coatings* (Chapman and Hall, New York).

21. Schneider, J. M., Rohde, S., Sproul, W. D., and Matthews, A. (2000), Recent developments in plasma assisted physical vapour deposition, *J. Phys. D: Appl. Phys.*, **33**, pp. 173–186.

22. Axén, N., Engqvist, H., Lööf, J., Thomsen P., and Hermansson, L. (2005), In vivo hydrating calcium aluminate coatings for anchoring of metal implants in bone, *Key Eng. Mater.*, **284–286**, pp. 831–834.

23. Doxa AB info, Techn Info (2004).

Chapter 11

Carriers for Drug Delivery Based on Nanostructural Chemically Bonded Bioceramics

This chapter presents the possibility of using chemically bonded bioceramics (CBBCs) as carriers for medicaments.

11.1 Chemically Bonded Bioceramics as Carriers for Drug Delivery: Introduction

Carrier materials for drug delivery of pharmaceuticals are based on a broad range of materials, such as organic polymers, metals, and sintered ceramics. General aspects of ceramics for use in drug delivery are given by Ravaglioli and Krajewski [1] and Lasserre and Bajpaj [2] and aspects of nanophase ceramics for improved drug delivery by Yang, Sheldon, and Webster [3].

There is a need for a carrier material for drug delivery that exhibits well-controlled microstructures, which lend the carrier material opportunities for selected and *well-controlled release* of the medicament. Issues regarding how and when the medicament is incorporated, and where and when it is released, are the theme of this chapter to ensure high delivery safety for medicaments with regard to the release pattern as well as safety aspects of the loaded carrier from chemical and mechanical aspects. A controlled carrier

Nanostructural Bioceramics: Advances in Chemically Bonded Ceramics
Leif Hermansson
Copyright © 2015 Pan Stanford Publishing Pte. Ltd.
ISBN 978-981-4463-43-0 (Hardcover), 978-981-4463-44-7 (eBook)
www.panstanford.com

material working also at the same time as a biomaterial meeting the above-mentioned criteria must also take account of and control of the setting and curing reactions in vitro and in vivo, as well as control of the porosity of the finally cured material and use of additives and processing agents to ensure an optimal microstructure.

This chapter deals with carrier materials based on chemically bonded bioceramics (CBBCs) primarily Ca-aluminates (CAs) and to some extent Ca-silicates (CSs). The property profile of these CBBCs and specifically their microstructures give these materials potential as carriers for drug delivery. Chemically bonded ceramics based on CA and CS phases as biomaterials have been presented in earlier chapters, especially Chapters 1–3. Geopolymers (metakaolin-based materials) are discussed in the literature also as carrier materials for medicaments [4, 5].

Below is described in some detail the CA carrier system with regard to technology and chemistry, biocompatibility, and specifically the microstructure and related loading possibilities of drugs in the carrier material. The development of microstructures includes different types of porosity, the amount of porosity, pore size and pore channel size, and combinations of different porosity structures. Specific surface area measurements (Brunauer–Emmett–Teller [BET]) of dried, fully hydrated CA yield BET values of >400 m^2/g, corresponding to a hydrate size of approximately 25 nm, and pore channel sizes of 1–5 nm, in accordance with values from the high-resolution transmission electron microscopy (HRTEM) analysis. Complementary porosity above 10 nm can be obtained by partial hydration of the precursor material or excess of water in the hydration step and pore sizes >100 nm by inert ceramic fillers with phases of oxides of Ti, Si, Ba, or Zr, the latter phases selected also to increase strength and radio-opacity of the carrier systems discussed. For more details, see Chapter 3 and Chapters 6–8 and Refs. [6–12]. The carrier material can be applied as a solid or a suspension for different types of intake. The drug carrier can also work as an injectable implant.

11.2 Important Aspects of Carriers for Drug Delivery

This section will to some extent treat and discuss the following topics of high importance when selecting materials to be used as carriers for drug delivery, namely:

- the chemical composition of the ceramic carrier material;
- the microstructure, including porosity of the carrier material;
- the optional use of inert additives; and
- the type of medicament and safety aspects.

The great freedom to select different ways of introducing drugs into the carrier system, and how and when release of the drugs can be executed in the CBBC system, is the topic of discussion in this chapter.

11.2.1 General Aspects

A couple of unique reaction conditions related to the production of materials yield the CBBCs a variety of microstructures, including porosity. These include microstructures having (1) different types of porosity, (2) different amounts of porosity, (3) different pores and pore channel sizes, and (4) combinations of different porosity structures.

Porosity generated during the hydration of CAs and CSs is open porosity due to the reaction mechanism and can be in the interval of 5–60 vol.%. The average pore channel size (i.e., the diameter of the pores formed between the particles of the hydrated material) may be 1–10 nm. The crystal size of the reacted hydrates is approximately 10–50 nm. This was established by BET measurements, where the specific surface area of dried hydrated CA was determined to be in the interval of 400–500 m^2/g, corresponding to a particle size of approximately 25 nm. When a short hydration time and/or a low amount of water, or moisture at relative humidity >70 %, are used, additional porosity is achieved with pore sizes in the interval of 0.1–1 μm due to an incomplete reaction.

The precursor powder cures as a result of hydration reactions, between the ceramic oxide powder, primarily CAs and/or CSs, and water. Through the hydration, new phases of hydrates are formed (crystalline and/or amorphous ones), which to a great part establish the microstructures needed to control the release of the drug incorporated in the material. The hydration mechanism of these systems involves a reaction where the total volume of the precursor materials and the water (solution) is reduced. This allows a carrier to exhibit open porosity throughout its body, even if the total porosity achieved is as low as approximately 5%–10%—a unique feature. The water-to-cement (w/c) ratio may be in the interval of 0.3–0.8. A ratio in the interval if 0.4–0.5 is near complete hydration of the CA material without any excess of water. Excess or limited amount of water favours complementary porosity, as does hydration at high relative humidity.

The setting time should be relatively short, below 30 minutes, and suitably in the interval of 5–15 minutes. The curing time and temperature are selected to produce a controlled microstructure. The carrier materials are suitably hydrated at a temperature above 30°C, since this yields the stable hydrates in the material and thus a more stable material. The curing before loading and/or before introduction of the material into the body can be done in water or in an environment with high relative humidity (>70 %). The setting and curing times and temperatures are of specific relevance when the carrier also works as an implant material.

Protection of the carrier and drug during passage through the body or life time in the body can be improved by coating of the precursor material. The coating may suitably be an acid-resistant or a hydrophobic layer. For medical agents sensitive to pH, the pH should be controlled in order to maintain their activity. A suitable pH is in the interval of 6–8. This can be achieved by introduction of a buffer. The buffer may favourably be a biocompatible one based on hydrogen phosphates.

The CBBCs selected yield the carriers a radio-opacity of approximately 1.5 mm. To impart a higher radio-opacity of the carrier, phases with high electron density are added. This allows the drug to be located in the body using X-ray techniques. Examples of such phases are ZrO_2 and Sr- and Ba-containing glasses.

The microstructure developed consists of nanosize hydrates and nanosize channels, located between said hydrates, having a size of about 10–50 nm and 1–5 nm, respectively. Complementary porosity above 10 nm can be achieved by (1) partial hydration of the precursor material, (2) excess of water in the hydration step, and/or (3) additional porous inert fillers, additional ceramics (such as hard particles or other hydrated phases), and other porous materials such as stable polymers and stable metals. The pore size can thus be controlled in a carrier from 1–2 nm to a micrometre-size level, typically <50 μm.

The property profile of the CA material can be seen in Table 11.1, which summarises the typical interval for each property. The intervals presented are mainly related to the completeness of the hydration, the water-to-cement ratio, and the type of filler particles introduced.

Table 11.1   Typical property values and intervals for CA-material-based systems

Property	Typical value	Interval
Compression strength (MPa)	150	60–270
Young's modulus (GPa)	15	10–20
Thermal conductivity (W/mK)	0.8	0.7–0.9
Thermal expansion (ppm/K)	9.5	9–10
Flexural strength (MPa)	50	20–80
Fracture toughness (MPam$^{1/2}$)	0.5	0.3–0.8
Corrosion resistance, water jet impinging, reduction (mm)	<0.01	–
Radio-opacity (mm)	1.5	1.4–2.5
Process temperature (°C)	>30	30–70
Working time (min)	3	<4
Setting time (min)	5	4–7
Curing time (min)	20	10–60
Porosity after final hydration (%)	15	5–60

11.2.2   Drug-Loading and Manufacturing Aspects

The loading of a drug can be performed in several ways [13]. The drug may be included, either partially or fully, in a powder or in a hydration liquid with or without any processing agents. The powder may be composed of non-hydrated or hydrated ceramics, porous additives such as sintered ceramics, stable polymers, or metals. The drug may be included in one or more of these powders or the liquid and may be mixed with or incorporated into any open porosity of the components.

The time and temperature for hydration are selected with regard to the drug and drug loading and to the selected release criteria.

Temperature, and the type of precursor powder, the amount of precursor powder, and processing agents control the time selected for manufacturing the carrier. The manufacturing of the carrier can be done completely before or during loading of the drug. This renders a controlled release time to be selected, from a few hours to days and months.

The drug is introduced in the carrier by mixing the drug into the precursor powder or the hydrated CBCs or other porous phases. The material can be formed into a paste by mixing it with a water-based hydration liquid. The powder can also be pressed into pellets, which thereafter are soaked in the liquid. The paste or the soaked pellets start to develop the microstructure, which to a great extent will contribute to the controlled release of the drug. The time and temperature after the mixing will determine the degree of hydration, that is, the porosity obtained. The porosity can be controlled within a broad interval of open porosity.

The drugs can be loaded in the water–liquid, in the pore system of inert filler particles, and in processing agents (accelerators, retarders, viscosity controlling agents, and other rheological agents). Thus drugs can be loaded both during formation of hydrates or after hydration by infiltration. The infiltration comprises water penetration of precursor materials or hydrated materials using wetting at normal pressure, during vacuum, or at overpressure. For hydrophobic medical agents, the agent can be easily mixed into the precursor powder or together with the ceramic or other filler materials.

The amount of drug loaded in the carrier is determined by the content of the drug in the dry powder and the hydration liquid. The liquids involved in the introduction of the drug into either the dry powder or the hydration liquid are easily controlled in charged amounts of liquids. The reaction takes preferentially place in high humidity, where no liquid in the carrier is vanished into the environment.

Optionally, the carrier powder may comprise inert oxides of Ti, Si, Ba, or Zr to increase strength or radio-opacity. The oxides may take the form of porous and/or dense particles. The incorporation of the drug or medical agent into the carrier material in the porous inert ceramic material may be performed by filling the pores of the inert ceramic with the drug, mixing it with the powder prior to mixing it with the hydration liquid, or mixing it with the hydration liquid prior to mixing it with the precursor powder. Depending on the type of drug delivery for which the carrier material is intended, a combination of one or more of these techniques may be used.

The carrier material may further comprise a third type of ceramics, including one or more of other hydrated or non-hydrated hydraulic phases, such as calcium phosphates, calcium sulphates, and hydroxyapatite. The carrier material may further comprise a fourth inert material of a porous polymer or porous metal. The carrier-loading capacity is estimated to be below 0.5 vol.% to as high as approx. 10 vol.%.

11.2.3   Drug Release Control Aspects

The following properties are of significance with regard to the carrier for controlling drug release: type of ceramic precursor for producing the chemically bonded ceramic, grain size distribution of the precursor powder particles and general microstructure of the material, the microstructure of the additional particles for drug incorporation, and additives to ensure complementary porosity.

11.2.3.1   Types of chemically bonded ceramics

The preferred chemical compositions, with an inherent property profile to meet the features described in this chapter, are those based on chemically bonded ceramics, which during hydration consume a controlled amount of water. The preferred systems available are those based on aluminates and silicates, which both consume a great amount of water. Phases such CA_2, CA, C_3A, and $C_{12}A_7$ and C_2S and C_3S in crystalline or amorphous state ($C = CaO$, $A = Al_2O_3$, $S = SiO_2$) may be used. The aluminate and silicate phases were synthesised at temperatures close to 1400°C for three to five hours. The CA and CS phases may be used as separate phases or as mixtures of phases. The above-mentioned phases, all in non-hydrated form, act as the binder phase (the cement) in the carrier material when hydrated.

11.2.3.2   Grain size distribution

The grain size of the precursor powder particles should be below 20 μm—this to enhance hydration. The precursor material is transformed into a nanosize microstructure during hydration.

This reaction involves dissolution of the precursor material and repeated subsequent precipitation of nanosize hydrates in the water (solution) and upon the remaining non-hydrated precursor material. This reaction favourably continues until all precursor materials have been transformed or to a porosity determined by partial hydration using the selected time and temperature.

11.2.3.3 Microstructure of additional particles (additives) for drug incorporation

The microstructure of the complementary additives which are penetrated by/loaded with the active medical agent is primarily characterised by its porosity, which should be open porosity in the interval of 15–70 vol.%. The average pore size determined by Hg porosimetry is in the interval of 0.1–10 μm. This is a complementary additive microstructure to that of the main structure based on chemically bonded ceramics. Examples of such additives include inert and hard ceramics such as oxides and/or carbides and/or nitrides. These phases yield a carrier material having increased strength and chemical resistance. The additional particles with a pore size in the interval of 0.1–10 μm are introduced to speed up the release rate from slow release down to a release time of a few hours (<5 hr) and can favourably be used to be loaded with additional drugs for rapid release. Figure 11.1 summarises the pore size intervals obtainable with different ceramics used for drug delivery.

Log (pore size in nm)

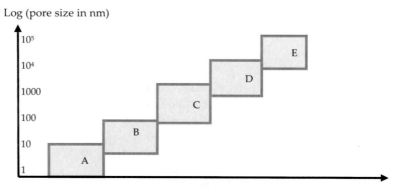

Figure 11.1 Pore size interval for different types of ceramics. A = fully hydrated, B = partially hydrated, C = sintered ceramics (submicron), D = sintered ceramics (partially or coarse grain), and E = others.

The amount of drug released in a specific case is controlled by the amount of the drug in the carrier and how the carrier material is designed. Using a mixed powder cement and an inert phase opens up for the use of combined drugs, for example, for rapid release based on the inert phase, medium release based on partially hydrated phases, and slow release based on fully hydrated cement phases. Test materials have been used with loading of the drugs in the interval of 1–10 mg/100 mg carrier material. Figure 11.2 shows schematically release rates using corresponding selected pore structures of the carrier material. The material types C, B, and A from Fig. 11.2 were used.

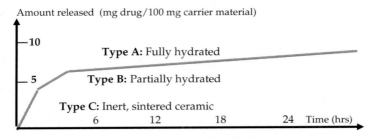

Figure 11.2 Schematic release of drugs using different pore structures.

11.2.3.4 Pharmaceutical compositions

The composition can be in the form of a solid or a suspension for different kinds of intake from oral intake to percutaneous injection.

The medicament can be of any kind. Preferable medicaments are those chosen from cancer/tumour treatment, vascular treatment, bone restoration, anti-bacterial and anti-inflammatory agents, pain relief drugs, anti-phlogistics, anti-fungal agents, anti-virus agents, analgesics, anti-convulsants, bronchodilators, anti-depressants, auto-immune disorder and immunological disease agents, hormonal agents, transforming growth factor beta (TGB-β), morphogenetic protein, trypsin inhibitor, osteocalcine, calcium-binding proteins (bone morphogenetic protein [BMP]), growth factors, bis-phosphonates, vitamins, hyperlipidaemia agents, sympathetic nervous stimulants, oral diabetes therapeutic drugs, oral carcinostatics, contrast materials, radio-pharmaceuticals, peptides, enzymes, vaccines and mineral trace elements, or other specific anti-disease agents.

Non-active ingredients can be added. Non-active ingredients mean water, alcohol, thickening agents, sweeteners, colours, antioxidants, or other additives which may be useful for stabilising the composition.

11.3 Conclusion and Summary

The ceramic carrier chemistry presented in this chapter allows for loading of almost any medicament.

The drugs can favourably be loaded in the water–liquid, in the pore system of inert filler particles, and in processing agents. Thus drugs can be loaded both during formation of hydrates or after hydration by infiltration.

For hydrophobic medical agents, the agent can be easily mixed into the precursor powder or together with the second ceramic filler.

The carrier material system described exhibits well-controlled microstructures on the nanosize level, which lends the carrier material opportunities for selected and controlled release of the medicament. The release time is controlled mainly by the contents of the hydrated chemically bonded cement phases—the higher the content of the cement, the longer the release time. The longest release time is achieved for fully hydrated phases with a water content close to the w/c required for complete hydration of the precursor CA.

By introducing optional additives, or by changing the w/c ratio, the release time can be controlled from a few hours to more than one day.

The release time is also dependent upon where the drug is placed. In cortical bone a release time of months seems possible.

The carrier may be used as a vehicle for transport and delivery of the medicament as an injectable implant. The combination of the material as a carrier and implant material makes site-specific placement of drugs and implants possible.

Acknowledgement

The author expresses his great gratitude to all personnel at Doxa AB for input under a ten-year period.

References

1. Ravaglioli, A. and Krajewski. A. (2000), *J. Mater. Sci. Mater. Med.*, **11**(12), pp.763–767.

2. Lasserre, A. and Bajpaj, P.K. (1998), Critical reviews, *Ther. Drug Carrier Sys.*, **15**, pp. 121–126.

3. Yang, L., Sheldon, B., and Webster, T. J. (2010), Nanophase ceramics for improved drug delivery, *Am. Ceram Soc. Bull.*, **80**(2), pp. 24–31.

4. Forsgren, J. (2010), PhD thesis, Uppsala University (Acta Universitatis Upsaliensis).

5. Duxson, P., Fernandez-Jimenez, A., Provis, J. L., Lukey, G. C., Palomo, A., and van Deventer, J. S. J. (2007), Geopolymer technology: current state of the art, *J. Mater. Sci.*, **42**(9), pp. 2917–2933.

6. Hermansson, L. (2011), Nanostructural chemically bonded Ca-aluminate based biomaterials, in *Biomaterials: Physics and Chemistry*, Ed. R. Pignatello (INTECH, Rijeka).

7. Adolfsson, E. (1993), *Phase and Porosity Development in the CaO-Al$_2$O$_3$-H$_2$O System*, Materials Science thesis, Uppsala University.

8. Hermansson, L., and Engqvist, H. (2006), Formation of nano-sized apatite coatings on chemically bonded ceramics, *Ceram. Trans.*, **172**, pp. 199–206.

9. Hermansson, L., Kraft, L., Lindqvist, K., Ahnfelt, N.-O., and Engqvist, H. (2008), Flexural strength measurement of ceramic dental restorative materials, *Key Eng. Mater.*, **361–363**, pp. 873–876.

10. Kraft, L. (2002), *Calcium Aluminate Based Cement as Dental Restorative Materials*, PhD thesis, Uppsala University.

11. Lööf, J. (2008), *Calcium-Aluminate as Biomaterial*, PhD thesis, Uppsala University.

12. Hermansson, L., Lööf, J., and Jarmar, T. (2009), Integration mechanisms towards hard tissue of Ca-aluminate based biomaterials, *Key Eng. Mater.*, **396–398**, pp. 183–186.

13. Hermansson, L. (2010), Chemically bonded bioceramic carrier systems for drug delivery, *Ceram. Eng. Sci. Proc.*, **31**, pp. 77–88.

Chapter 12

Clinical Observations and Testing

This chapter presents clinical aspects of chemically bonded bioceramics (CBBCs) and clinical testing of new biomaterials. The presentation will mainly be focused upon some Ca-aluminate-based biomaterials.

12.1   Clinical Evaluation: An Introduction

New biomaterials have to be thoroughly tested regarding critical general properties. This is performed according to international standards (International Organisation for Standardisation [ISO]) and described in Chapter 5. Before clinical evaluation starts the new biomaterials have to be thoroughly tested related to, among others, biocompatibility and, in case when appropriate, also aspects of bioactivity. Biocompatibility, including bioactivity, is treated in some detail in Chapters 7 and 8.

For most new biomaterial applications, clinical testing is required, and the evaluation differs depending on the specific application the biomaterial is intended for.

Below are presented as examples of clinical evaluation results from Ca-aluminate-based biomaterials intended for dental and orthopaedic use.

Nanostructural Bioceramics: Advances in Chemically Bonded Ceramics
Leif Hermansson
Copyright © 2015 Pan Stanford Publishing Pte. Ltd.
ISBN 978-981-4463-43-0 (Hardcover), 978-981-4463-44-7 (eBook)
www.panstanford.com

12.2 Dental Biomaterial Evaluation

12.2.1 Introduction

For minor changes of already approved biomaterials for specific applications, no clinical evaluation is required. However, it is always a safety for both producers and users of a biomaterial for a specific application to also rely on clinical testing.

In all clinical evaluation studies certain inclusion and exclusion criteria are used. Typically such criteria for dental applications are presented in Tables 12.1 and 12.2 below.

Table 12.1 Inclusion criteria

Criteria	Description
1	Written consent to participate in study must be obtained.
2	Patients must be at least 18 years of age.
3	Patients must require and be treatment-planned for at least one unit of fixed restorative dentistry.
4	Patients must be available for required recalls, as outlined in the protocol.
5	Patients selected for the study will have a low-to-moderate caries rate, a stable periodontal status with good home care and not involving extensive alveolar bone loss, and/or gingival recession with tooth mobility.
6	Teeth treated should be assessed clinically as vital (in selected situations, root-canal-treated, and non-vital teeth may be included in the study) without evidence of pulpal or surrounding periodontal soft-tissue pathology.
7	Patients must be in generally good health, with no medical contraindication to dental treatment.

Below are presented parts from two clinical testing procedures, one for a new dental luting cement for crowns and bridges and the other for an experimental endodontic retrograde filling material [1, 2].

Table 12.2 Exclusion criteria

Criteria	Description
1	Intended patients do not meet all inclusion criteria.
2	Patients have untreated periodontal disease.
3	Patients have pathologies or systemic problems which would not allow the dental procedures in this study to take place.
4	Patients have rampant caries.
5	Patients have severe bruxing or clenching of teeth.
6	Teeth are non-vital, exhibit pulpal pathologies, or have expected pulp exposures (exception when inclusion of a root-canal-treated, non-vital tooth is deemed warranted due to clinical circumstances or in the judgment of the investigators).
7	The subject is pregnant.

12.2.2 Dental Luting Cement: Prospective Observations

Long-term success of fixed restorations depends on a range of factors, including the quality of the luting agent used, biocompatibility, insolubility, and resistance against degradation, all of which help maintain the seal at the restoration margins, thus preventing ingress of bacteria leading to leakage, sensitivity, and secondary decay.

The progression of luting agents has evolved with a succession of chemistries over the past century or more, including zinc phosphate, polycarboxylate, glass ionomer, resin, resin-modified glass ionomer (RMGI), and self-adhesive resin cements [3–9]. These cement compositions are now challenged by a new hybrid calcium aluminate/glass ionomer cement, Ceramir® C&B (CM), originally named XeraCem (Doxa Dental AB, Uppsala, Sweden), a luting agent intended for permanent cementation of crowns and fixed partial dentures (FPDs), gold inlays and onlays, prefabricated metal and cast dowel and cores, and all-zirconia or all-alumina crowns, as well as Li-silicate-based crown materials [10]. The cement is a water-based composition comprising calcium aluminate and glass ionomer components and has been demonstrated to be bioactive. The term 'bioactivity' refers to a property of this new cement to form hydroxyapatite (HA) when immersed in vitro in a physiological phosphate buffered saline solution.

The introduction of any new cement chemistry necessitates assessment of its laboratory and clinical performance. The laboratory performance of this new cement has been assessed with respect to a number of performance criteria. Assessment of compressive strength, film thickness, and setting time all conformed favourability to the ISO standard for water-based luting cements [11].

Comparative in vitro microleakage performance of this new bioactive cement has also been assessed by two methodologies. Dye leakage analysis in cemented crowns concluded that the Ca-aluminate-based cement (Ceramir® C&B, CM) demonstrated significantly less leakage than a conventional glass ionomer cement, Ketac-Cem (KC). An in vitro bacterial leakage model comparison of CM to a conventional glass ionomer luting cement, KC, and a resin-modified glass ionomer cement (Rely X Luting Plus, RX) demonstrated that the groups cemented with CM and RX showed no significant difference in microleakage patterns ($p > 0.05$), while both recorded significantly lower microleakage scores ($p < 0.05$) than the group cemented with KC [12, 13].

Biocompatibility ranks as one of the most important properties of a final luting cement, and as such, a number of in vitro and in vivo tests (as recommended by ANSI/ADA Spec. 41 and ISO 10993) were conducted prior to the clinical investigation to evaluate the biocompatibility of Ceramir® C&B cement [14, 15].

Results of the Ames test for mutagenicity indicated that this new cement formulation did not induce genemutations. In vitro cytotoxicity testing indicated cell responses ranging from none to mildly cytotoxic, an acceptable response. The skin sensitisation test (in guinea pigs) indicated that this cement is not a skin sensitiser. Testing for mucous membrane irritation (hamster pouch test) indicated that it produced no local irritation. Pulpal testing in *Rhesus macaques*, according to ANSI/ADA Spec. 41, indicated a virtual absence of pulpal inflammation, at both 30- and 85-day evaluation periods, after CM was used to cement composite resin inlays in a class V preparation.

Retention is perhaps the most critical factor in the performance of a luting cement. A comparative, in vitro crown retention study was conducted (also prior to the clinical evaluation) to assess the retentive properties of the new cement [16, 17]. Results of this test indicated that it demonstrated retentive values equivalent (no statistically significant difference) to a self-adhesive resin cement, Rely X Unicem, but were significantly higher than those of a conventional glass ionomer (KC) and zinc phosphate cements. The cement is approved to be marketed in the United States in its powder–liquid, hand-mixed version and in a capsule delivery system. See Fig. 12.1 below.

Figure 12.1   The cement system for Ceramir® C&B.

In evaluating the dental cement, clinical study parameters used in terms of specific patient and restoration data are provided in Table 12.3.

Table 12.3   Baseline clinical study parameters

No. of patients	17
No. of restorations	38
Demographics	9 women, 8 men
Ages	25 to 79 years
No. of single-unit crowns	23
No. of FPDs/fixed splints	6 FPDs (13 abutments)/splint (2 units)
Vital/non-vital prepared teeth	31 vital/7 RCT-treated

The cement-handling parameters evaluated in the clinical study were dispensing, mixing, working time, seating characteristics, and ease of cement removal (see Table 12.4). Clinical measurement data is also detailed in Table 12.5 and consisted of pre- and post-cementation sensitivity according both to categorical and visual analogue scale (VAS)-based measurements, marginal integrity, marginal discoloration, and secondary caries according to the modified Ryge (United States Public Health Service [USPHS]) criteria [18]. Gingival response was evaluated pre- and post-cementation by means of the Loe and Silness gingival index (GI) [19]. Retention was assessed by the criteria described in Table 12.5. The procedure for categorical tooth sensitivity data was simply to ask the patient to characterise any pain or discomfort (or lack thereof) for the treated tooth/teeth in question on the basis of four possible choices: none, slight, moderate, and severe. The response was recorded. For the VAS assessment of tooth restoration sensitivity, the patient physically marked a point on a line of 100 mm (10 cm), with one end (left) of the line indicating 'no sensitivity or pain' and the other (right) end of the line indicating 'maximum sensitivity or pain'. The point marked by the patient was measured to the nearest millimetre and recorded at each designated time point.

Table 12.4 Measurement parameters for clinical study/cement measurement data

Cement measurement data	Description
Dispensing	Easy/difficult
Working time	OK/too short/too long
Mixing	Easy/difficult
Seating characteristics	Restoration completely seated after cementation: Yes/No
Ease of cement removal	Easy/normal/difficult

Table 12.5 Measurement of parameters for clinical study data

Clinical measurement data	Description
Sensitivity	Alpha: none Beta: slight Charlie: moderate Delta: severe
Retention	Restoration in place: Yes: alpha; No: delta
Marginal integrity	See Table 12.6

Clinical measurement data	Description
Caries	See Table 12.6
VAS sensitivity	Measurement by patient in millimetre on a continuous line between 'no' and 'extreme' pain or discomfort
GI index	0–1–2–3 scoring as per Loe and Silness

Table 12.6 Criteria for clinical evaluation (including criteria adopted from modified Ryge criteria and gingival inflammation index of Loe and Silness)

Theme	Description
Rating characteristic	Alpha/Bravo/Charlie/Delta
Marginal adaptation	Alpha: There is no visible evidence of a crevice along the margin which the explorer will penetrate. Bravo: There is visible evidence of a crevice along the margin which the explorer will penetrate. Charlie: The explorer penetrates a crevice, reaching dentin or base/core material. Delta: Restoration is mobile or missing.
Marginal discolouration	Alpha: There is no discolouration evident along visible marginal areas between the restoration and the tooth structure. Bravo: Discolouration is present but superficial (has not penetrated along the margin in a pulpal direction). Charlie: Discolouration is present and has penetrated along the margin in a pulpal direction. Delta: N/A.
Secondary caries	Alpha: No caries as evidenced by softness, opacity, or evidence of demineralisation at the margin of the restoration Bravo: Evidence of caries at the margin of the restoration Charlie: N/A Delta: N/A

(*Continued*)

Table 12.6 (*Continued*)

Theme	Description
Pre- and post-operative sensitivity	Alpha: No sensitivity Bravo: Slight sensitivity Charlie: Moderate sensitivity Delta: Severe sensitivity
Retention	Alpha: Restoration still in place Bravo: N/A Charlie: N/A Delta: Restoration not in place
Gingival inflammation	0: Absence of inflammation 1: Mild inflammation—slight change in gingival colour and little change in texture 2: Moderate inflammation—moderate glazing, redness, oedema, and hypertrophy; bleeding or pressure at entrance to sulcus 3: Severe inflammation—marked redness and hypertrophy; tendency to spontaneous bleeding or ulceration

Summary results of the clinical evaluation are presented in Table 12.7.

Table 12.7 Clinical data for modified Ryge criteria, GI, and VAS scores

	Baseline	1 month	6 months	1 year	2 years
No. of patients recalled	17	17	17	15	13
No. of restorations/ Abutments	38	38	38	31	27
% Alpha: absence of Caries	100	100	100	100	100
% Alpha: marginal Integrity	100	100	100	100	100
% Alpha: marginal discoloration	100	100	100	100	100
Average VAS score (mm)	7.6	3.1	0.4	0.2	0.0
Average GI	0.56	0.10	0.11	0.16	0.21

In Fig. 12.2 below is shown four crowns (porcelain on metal) cemented by the Ceramir C&B Ca-aluminate based dental cement.

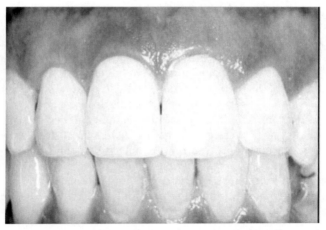

Figure 12.2 Four dental crowns cemented by Ceramir® C&B.

12.2.3 Endodontic Fillings: A Retrospective Investigation of a Ca-Aluminate-Based Material in Root Canal Sealing

Calcium aluminate hydrate (CAH) belongs to the same material group as MTA (ProRoot) and Biodentine based on Ca-śilicates, chemically bonded ceramics [20–22]. General properties of the materials are described in Chapters 3–5 and Chapters 7–9.

The objective of this study was to conduct a long-term clinical study using a Ca-aluminate-based material as a canal sealer and as a retrograde filling material. The study was conducted on human volunteers who had signed an informed consent form. A total of 17 patients were treated, who were diagnosed with either a chronic perapical lesion or who were in need of a retrograde filling in failed endodontically treated teeth. A total of 8 orthograde and 14 retrograde filling treatments were performed. Recall visits were scheduled after two and five years. The results are assembled from the work of Dr Polhagen, different remitting dentists, and Dr Mikael Saksi [2].

The orthograde root canal treatments were carried out according to standardised established techniques, using NiTi files and

NaOCl. The material was mixed with a solvent into an appropriate consistency and was then put into a syringe and injected in a thin cannula or with capillary tips into the root canals, where it subsequently was condensed with coarse gutta-percha points. For the retrograde root fillings (rf's) a modern technique with a dental operating microscope was used. A conventional surgery procedure, involving laying of flaps, removal of tissues from outside the root canal space, including bone, periodontal membrane, and periostum, was conducted. The apex was detected with a surgery microscope and rinsed and prepared with an ultrasonic device. The material was then inserted and condensed with dental instruments. After therapy the patients' teeth were examined and investigated with X-ray, and after two years or more the therapy success of the treatment was evaluated from patient inspection and from new X-ray pictures. Three questions regarding subjective symptoms also were put to the patients for consistent judgement of their subjective experience of the therapy:

1. Have you had any persistent symptoms?
2. Do you know which tooth was treated?
3. Can you feel any symptoms at the tooth apex?

The results of both the clinical examination and the subjective symptoms were graded into different groups related to the success of the therapy, as presented in Table 12.8.

Table 12.8   Success rate description

Group	Observation	Description
Score 1	Complete healing	No subjective symptoms at all and no apical radiolucency at X-ray control
Score 2	Incomplete healing	No subjective symptoms at control but having symptoms occasionally or only reduced apical radiolucency at X-ray control
Score 3	Uncertain	No subjective symptoms at control but having symptoms occasionally and only reduced apical radiolucency at X-ray control
Score 4	Failure	Symptoms present and/or no change of apical radiolucency

For 13 of the 17 treated patients the diagnosis was chronic perapical osteitis (cpo). They were treated with retrograde rf therapy. Three patients suffered from trauma or chronic perapical destruction, and these patients were treated by orthograde therapy. Two patients had repetitive retrograde therapy. Out of 17 patients (22 teeth) treated, 16 patients (21 teeth) were examined with follow-up X-ray after treatment and also after two years or later. The additional patient was asked about symptoms. Score 1 or 2 is considered successful, and score 3 or 4 is considered a failure. From the clinical examinations and subjective symptoms 18 teeth had complete healing, 3 teeth had incomplete healing, and 1 tooth was not healed. Of the three patients with score 2, one patient (#4) occasionally felt symptoms, one patient (#6) had no symptoms according to a phone call conversation but was never examined in the clinic, and one patient (#17) was without symptoms but the periodontal bone had not fully been recovered. Patient 5 with unacceptable results had the distal root (dr) canal healed, but the other mesial root (mr) filling had failed and still caused distress for the patient. After five years some of the original patients could not be reached. The whole study with all patient numbers is summarised in Table 12.9.

Table 12.9 Summary of the study

Patient no.	Diagnosis, indications, surgical procedure	Symptoms at control, group score
1	11 cpo. Trauma in frontal maxilla, with swelling. Open apex. OG[a]	2 and 5 years, healed No symptoms, 1
2 a b c	22, 21, 12 all with cpo. Not fully healed after trauma and OG treatment. RG[b]	2 and 5 years, healed, 1
3	cpo 16. Patient with pain, difficult to diagnose. Patient informed of a dubious prognosis for her pain treatment. RG	2 and 5 years, completely healed No pain resisting, 1
4	36 juxtradicular, cpo, possible crack. RG, distal root amputation	2 years, completely healed, No 5-year control, 2

(Continued)

Table 12.9 (*Continued*)

Patient no.	Diagnosis, indications, surgical procedure	Symptoms at control, group score
5	cpo 26, mb and db roots. RG	2 years, symptoms remained in mb; db healed, After 5 years, no problems, 4 (2)
6	cpo 15, RG	No X-ray control at 2 years, 5 years, X-ray control, symptom free, 2
7	Exacerbations, cpo 21, apical symptoms. RG	2 years, healed, no symptoms, 1
8	Chronic perapical destruction 11. OG	2 years, completely healed, – 1
9	cpo 11, RG	2 years, completely healed Apical parestesi remaining, 5 years, no symptoms, 1
10	cpo 21, retrograde amalgam rf removed with ultra-sonic. Repetitive RG therapy	2 and 5 years, healed, 1
11	cpo 23, RG	2 years, healed, –, 1
12 a b c d	cpo 12 11 21 22 All RG	2.5 and 5 years X-ray Every tooth healed, 1
13	cpo 13, RG	2.5 and 5 years healed, 1
14	cpo 21, trauma OG with CA material on the apical third of the tooth	2 years healed, 1 –
15	cpo 21, RG	2.5 years healed, – 1
16	cpo 11, exacerbations, RG	2 and 5 years healed, 1
17	cpo 21 Repetitive RG therapy	2.5 years X-ray. The apical radiolucency reduced. Incomplete healing, without periodontal contour, 5 years symptom-free 2 (1)

[a]OG = orthograde rf
[b]RG = retrograde rf

The X-ray examination revealed good results, and selected pictures at the two-year control and at the five-year control are presented in Figs. 12.3 and 12.4.

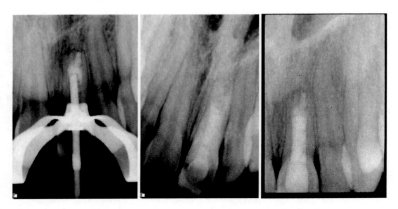

Figure 12.3 Patient 14, tooth 21. Orthograde condensing with a gutta-percha point (left), after treatment (middle), and at the two-year control (right).

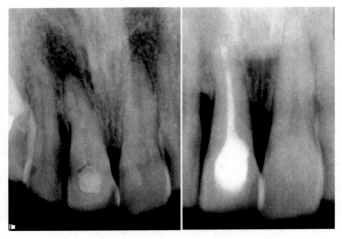

Figure 12.4 Patient 9, tooth 12. Before treatment (left) and at five-year control after treatment (right).

12.3   Orthopaedic Biomaterial Evaluation

12.3.1   Introduction

Nanostructural chemically bonded bioceramics (CBBCs) are found only within a few orthopaedic applications so far. Below are presented some studies where Ca-aluminate-based biomaterials have been tested within orthopaedics.

Osteoporosis is by far the biggest cause of fractures. It affects an estimated 75 million people only in Europe, the US, and Japan, and the global prevalence is forecast to grow quickly. One in three women that have passed menopause will experience osteoporotic fractures, as will one in five men. There are an estimated two million osteoporotic fractures every year in the US and Europe. Vertebral compression fractures (VCFs) caused by low-energy trauma are increasingly common in osteoporotic patients. For patients with severe osteoporosis, simple daily life activities are enough to cause painful collapse of one or several vertebrae. In individuals with less pronounced osteoporosis, a low-energy type of loading such as falling may cause compression fractures. The back pain lasts approximately 4–10 weeks, but in some cases it may persist for months with the risk of becoming chronic and causing other complications such as loss of lung capacity, decreased appetite, and even depression. Conventional treatment includes bed rest, bracing, and analgesics for pain relief. However, reduced physical activity accelerates bone loss, whereby this treatment may further increase osteoporosis. Current treatments for VCF are focused on vertebral body augmentation, that is, strengthening of the vertebral body by injecting bone cement. Augmentation also potentially includes therapies aimed at the prevention of vertebral fractures and would involve the delivery of biologically active compounds in order to increase bone mass. The VCF market is dominated by two procedures, percutaneous vertebroplasty (PVP) and kypho-vertebroplasty (KVP). These are described in Chapter 10.

Poly(methyl methacrylate) (PMMA) bone cement is currently the dominant commercially available material for vertebroplasty under various brand names, and it has been extensively characterised in the literature [23, 24]. Although PMMA is recognised as a successful filler material, there are shared concerns in the field that it is not the optimal material, for several reasons. Disadvantages of PMMA have been identified as:

(1) no biologic potential to remodel or integrate into the surrounding bone;

(2) no direct bone apposition;

(3) high polymerisation temperature;

(4) potential monomer toxicity; and

(5) low radio-opacity.

With CBBCs (phosphates, sulphates, and aluminates) [25–34] there is a significant opportunity for a VCF cement to exhibit:

(1) high radio-opacity;

(2) high initial and predictable viscosity to improve safety during procedure;

(3) optimised injection characteristics (easily injected with minimal risk of leakage);

(4) good integration with vertebral bone;

(5) hardening at reduced temperature;

(6) immediate stability;

(7) mechanical strength adapted for early and active rehabilitation; and

(8) a possibility to deliver active compounds.

12.3.2 Clinical Studies

Below are, as examples of clinical evaluation, presented studies in vertebroplasty using the PVP technique. In the study a Ca-aluminate-based material was used. The inclusion and exclusion criteria for these studies are shown in Tables 12.10 and 12.11.

Table 12.10 Inclusion criteria

Criteria	Description
1	Males and females, 50 years or older
2	Radiographic evidence of 1 to 3 low-energy VCFs between T5 and L5 with < or = 70% height loss of the vertebral body
3	Back pain corresponding to the level of at least one of the VCFs

4	Pain corresponding to at least 5 (50) out of 10 (100) on a VAS scale
5	Oedema in the vertebra as seen at MRI
6	Substantial requirement of analgesics
7	Signed written informed consent

Abbreviation: MRI, magnetic resonance imaging.

Table 12.11 Exclusion criteria

Criteria	Description
1	Previous spinal surgery
2	Neurological deficit or radiculopathy associated with the level to be treated
3	Complete vertebral compression/collapse
4	Disabling back pain secondary to causes other than the acute fracture
5	Malignancy
6	Above reference value of serum creatinine
7	Uncorrectable coagulation disorder
8	Immunological deficit
9	Active infections of any kind
10	MRI contraindication
11	Participation in other clinical studies within the last 30 days
12	Other clinical conditions that makes the patient unsuitable for PVP
13	Disruptions and other kinds of fractures in the area under study

A complete study comprises several aspects related to baseline actions, imaging, questionnaires, safety, surgery, and laboratory work. A time table of the overall study is presented in Table 12.12.

Descriptions of tools used in the evaluation are also described in more detail below.

Visual Analogue Scale

Pain relief was measured using a VAS graded from 0 to 10 (0–100 mm) [18]. The principal investigator (PI) was to ask the patient, *How*

Table 12.12 Schedule of events indicating the procedures performed at each study visit/telephone contact

	Screening	Procedure	24 hours	1 week	1 month	3 months	6 months	12 months
Informed consent form	X							
Medical history	X							
Demography	X							
Radiographs		X						
CT			X					
MRI	X							X
VAS	X		X	X	X	X	X	X
Oswestry	X		X	X	X	X	X	X
Mobility	X		X	X	X	X	X	X
QoL	X					X	X	X
Use of analgesics	X		X	X	X	X	X	X
Safety*								
A			X	X	X	X	X	X
B	X		X	X	X	X	X	X
Surgery*								
C		X						
D		X						
E		X						
Laboratory*								
F	X							X

Abbreviations: CT, computed tomography; QoL, quality of life; TSH, thyroid-stimulating hormone.

*Safety: Surgery and Laboratory: A = adverse event/SAE/complications; B = concomitant medications; C = ECG supervision; D = oxygen saturation; E = handling characteristics; F = coagulation, kidney, TSH

severe is your pain today? Please place a vertical mark on the line to indicate how bad you feel your pain is today.

At all visits to the clinic, the patient was to place a vertical mark on the VAS himself or herself. However, at the one-week assessment over the telephone, the PI or the study nurse was to mark a line corresponding to the amount of pain graded by the patient.

Change from screening for VAS was to be measured at all follow-up visits/contacts.

Mobility Score

Mobility before and after PVP was to be graded from 0 to 3 by giving one of four alternative answers:

0. No impairment
1. Walking with help/assistance
2. Wheelchair
3. Bedridden

Mobility was to be graded at screening and at all follow-up visits/contacts.

Oswestry Back Index

The validated Oswestry back index (form BI-100) licensed from ACN Group Inc. was used to assess how the patient's back condition affects his or her daily life. Translations of the Oswestry back index to local languages (German and Italian in this study) were made by an authorised translator.

The Oswestry back index questionnaire aims to give information on how the patient's back condition affects his/her everyday life. The Oswestry back index contains 10 questions with 6 alternative answers (0–5) regarding pain and everyday activities. The questions were to be answered by choosing the 'best answer' describing the 'typical' pain and/or limitations within the last week or two. Only one answer was to be chosen.

The questionnaire was to be performed by the PI or the study nurse but preferably by the same person at every visit. The points were to be added up, divided by 50, and multiplied by 100 to give the percent disability.

The Oswestry back index score interpretation is as follows:

- 0%–20%, minimal disability: The patient can cope with most living activities. Usually no treatment is indicated apart from advice on lifting, sitting, and exercise.

- 21%–40%, moderate disability: The patient experiences more pain and difficulty with sitting, lifting, and standing. Travel and social life are more difficult, and the patient may be disabled from work. Personal care, sexual activity, and sleeping are not grossly affected, and the patient can usually be managed by conservative means.
- 41%–60%, severe disability: Pain remains the main problem in this group, but activities of daily living are affected. These patients require a detailed investigation.
- 61%–80%, crippled: Back pain impinges on all aspects of the patient's life. Positive intervention is required.
- 81%–100%: These patients are either bed-bound or exaggerating their symptoms.

Quality of Life

QoL was to be assessed by the SF-12v2™ Health Survey in local languages, SF-12v2™ Health Survey Standard, Germany (German), and SF-12v2™ Standard, Italy (Italian) Version 2.0, 8/03 licensed from QualityMetric Inc. and Medical Outcomes Trust.

The SF-12® Health Survey is a multipurpose short form with 12 questions selected from the SF-36® Health Survey. In this study, the SF-12v2™ Health Survey was applied. In addition to scoring summary measures, an eight-scale health profile was to be scored related to Physical Functioning (PF), Role Physical (RF), Bodily Pain (BP), General Health (GH), Vitality (V), Social Functioning (SF), Role Emotional (RE), Mental Health (MH).

A four-week recall period was to be used to capture a representative and reproducible sample of recent health, not unduly affected by daily or momentary fluctuations. The questionnaire was to be performed by the PI or the study nurse but preferably by the same person at every visit.

12.3.2.1 A prospective clinical study in PVP

The example of a clinical study within orthopaedics presented below relates to a Ca-aluminate-based biomaterial (Xeraspine) used in PVP. PVP is a minimally invasive technique in which medical-grade cement is injected into compressed fractures in order to relieve pain and provide stability. A large number of studies have shown that this treatment provides good pain relief and improved mobility with a lasting effect for several years.

The conclusion from a pilot study in Germany of Xeraspine® in PVP on eight patients with VCF was that injection with Xeraspine® is a safe treatment of VCF which offers substantial and durable pain relief. Even with the promising pilot study results, data from a larger study was considered necessary to deepen the knowledge of the product and to gain more experience of Xeraspine® as an injectable bioceramic.

In the open, non-comparative study, the subject had 1–3 vertebrae treated with Xeraspine®. To evaluate the outcome of the procedure, pain relief was measured using VAS, whereas mobility improvement was measured by four three-graded mobility questions. The Oswestry disability questionnaire and the SF-12v2™ QoL questionnaire were also used for evaluation. The aim was to follow the patients for 12 months after the procedure. A general description of the study is presented in Table 12.13.

Table 12.13 PVP clinical test

Clinical study	Description
Clinics/patients	7 European clinics (radiologists); a total of approximately 50 patients in the study
Study design	Open, multicenter, prospective, non-comparative clinical study with a follow-up period of 12 months
Primary objective	To measure pain at rest in patients with low-energy vertebral compression fracture before and after PVP treatment using pain improvement, pre-operative vs. 24-hour (primary end point), 1-week, 1-, 3-, 6-, and 12-month post-operative, evaluated with VAS scores
Secondary objectives	To measure clinical, radiological, and physical/chemical variables of importance for establishing further effectiveness and safety

12.3.3   Presentation of clinical results

Below are presented as examples some results from the clinical evaluation of the PVP study described in Table 12.13.

12.3.3.1   Primary effectiveness variable

Visual Analogue Scale
The amount of back pain at rest corresponding to the level of VCF was to be rated by the patient by placing a vertical mark on a 100 mm VAS with the extremes 'No pain' (0) and 'Worst possible pain' (100).

At screening, the patients presented back pain within an interval of 2–100 mm VAS and a mean of 73.8 mm VAS. Already within 24 hours after Xeraspine® PVP treatment, the mean VAS at rest was reduced to 28.8. A slight additional pain reduction could be seen at one week after the procedure, and it remains unchanged throughout the three months studied (see Fig. 12.5 below).

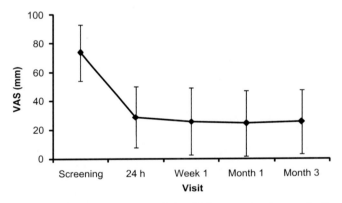

Figure 12.5   Pain at rest measured by VAS at screening and after the procedure.

At 6 and 12 months the VAS results were in the same range as for month 3. See Fig. 12.6.

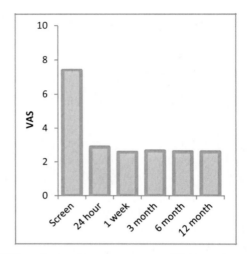

Figure 12.6   PVP study results up to one year (VAS in cm).

Corresponding results from a KVP study are shown in Fig. 12.7.

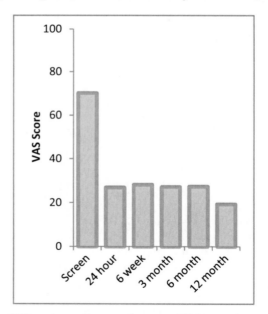

Figure 12.7   KVP study results up to one year (VAS in mm).

12.3.3.2 Secondary effectiveness variables

Mobility Improvement
At screening, 16 patients were walking with assistance, 1 was in a wheelchair, and 3 were bedridden. Thirty-two patients reported 'no impairment' at screening. Of the 52 patients with 3-month data, 37 patients reported no change in mobility 24 hours after the procedure, whereas 2 out of 3 previously bedridden patients and 1 patient in a wheelchair were walking with assistance. The patient who remained bedridden at 24 hours was able to walk with assistance 1 week after the procedure.

Quality of Life
The SF-12v2™ Health Survey score ranges from 0 to 100, where 0 symbolises the poorest health and 100 is the best health. In the 46 patients who had no missing values at screening or at the three-month evaluation, the SF-12v2™ total score increased from 36.3 at the screening assessment to 60.8 three months after the procedure. The mean difference between the screening and three-month scores was 21.7, which demonstrates a significant improvement in general health three months after PVP treatment with Xeraspine®. See Table 12.14 below.

Table 12.14   SF-12v2™ total score between screening and a three-month visit

	Screening	3 months
No. of patients	54	50
Missing patients in some part of the studies	5	9
Mean value	36.3	60.8
Standard deviation	18.1	22.9
Min. value	0	13
Median	37.3	66.3
Max. value	77	94

The SF-12v2™ scale items were aggregated to give an eight-scale health survey profile. The eight scales are Physical Functioning (PF), Role Physical (RP), Bodily Pain (BP), General Health (GH), Vitality (V), Social Functioning (SF), Role Emotional (RE), and Mental Health (MH). The mean difference between screening and the three-month assessment for each of the eight scales may differ (see Table 12.15).

Table 12.15 Aggregated eight-scale health survey, SF-12v2™ scores

Profile	Screening	3 months
PF		
No. of patients	57	51
Mean	28.9	53.4
Standard deviation	29.4	30.8
RP		
No. of patients	55	52
Mean	19.3	56.5
Standard deviation	19.7	28.5
BP		
No. of patients	57	52
Mean	22.8	59.1
Standard deviation	24.2	36.7
GH		
No. of patients	57	52
Mean	36.1	48.7
Standard deviation	25.9	29.4
V		
No. of patients	57	52
Mean	36.0	55.8
Standard deviation	24.1	23.5
SF		
No. of patients	57	52
Mean	51.8	74.0
Standard deviation	29.5	22.1
RE		
No. of patients	55	51
Mean	52.3	73.8
Standard deviation	34.5	26.1
MH		
No. of patients	57	52
Mean	43.9	61.3
Standard deviation	21.9	22.0

It stands clear that the patients experienced most improvement in the Role Physical and Bodily Pain categories. Role Physical is related to physical health and includes the SF-12v2™ questions 'Accomplished less than you would like' and 'Were limited in the kind of work or other activities'. Bodily Pain is represented by the question 'How much did pain interfere with your normal work (including both work outside the work and housework)?'.

12.4   Overall Conclusions

- New biomaterials have to be thoroughly tested regarding critical general properties. This is performed according to international standards (ISO). Before clinical evaluation starts the new biomaterials have to be thoroughly tested related to, among others, biocompatibility and, in case when appropriate, also aspects of bioactivity.

- For most new biomaterial applications, clinical testing is required, and the evaluation differs depending on the specific application the biomaterial is intended for. For minor changes of already approved biomaterials for specific applications, no clinical evaluation is required. However, it is always a safety for both producers and users of a biomaterial for a specific application to also rely on clinical testing.

- In all clinical evaluation studies certain inclusion and exclusion criteria are used. These are introduced to make the evaluation as safe as possible from, especially, the patient's viewpoint. This often leads to excellent results expressed as a high implant survival degree, but when the implants after some time come to a general use, the survival rate may be lower. The general discussion related to peri-implantitus is a sign of this.

- PMMA bone cement is currently the dominant commercially available material for vertebroplasty under various brand names. Although PMMA is recognised as a successful filler material, there are shared concerns in the field that it is not the optimal material, and PMMA-based materials are challenged by nanostructural CBBCs based on phosphates, sulphates, silicates, and aluminates.

Acknowledgement

The author thanks the personnel at Doxa Company, Sweden, and the Materials Science Department at Uppsala University for valuable inputs during a two-decade period.

References

1. Jefferies, S. R., Pameijer, C. H., Appleby, D. C., Boston, D., and Lööf, J. (2013), A bioactive dental luting cement: its retentive properties and 3-year clinical findings, *Compend. Contin. Educ. Dent.*, **34**(Spec No 1.), pp. 2–9.

2. Kraft, L., Saksi, M., Hermansson, L, and Pameijer, C. H. (2009), A five-year retrospective clinical study of a calcium-aluminate in retrograde endodontics, *J. Dent. Res.*, **88**(A), p. 1333.

3. Drago, C. (1996), Clinical and laboratory parameters in fixed prosthodontic treatment, *J. Prosthet. Dent.*, **76**, pp. 233–238.

4. Black, S. M., and Charlton, G. (1990), Survival of crowns and bridges related to luting cements, *Restor. Dent.*, **6**, pp. 26–30.

5. Pameijer, C. H., and Nilnert, K. (1994), Long term clinical evaluation of three luting materials, *Swed. Dent. J.*, **18**, pp. 59–67.

6. Kern, M., Kleimeier, B., Schaller, H. G., and Strub, J. R. (1996), Clinical comparison of postoperative sensitivity for a glass ionomer and a zinc phosphate luting cement, *J. Prosthet. Dent.*, **75**, pp. 159–162.

7. Jokstad, A., and Mjor, I. A. (1996), Ten years' clinical evaluation of three luting cements, *J. Dent.*, **24**, pp. 309–315.

8. Jokstad, A. (2004), A split-mouth randomized clinical trial of single crowns retained with resin-modified glass ionomer and zinc phosphate luting cements, *Int. J. Prosthodont.*, **17**, pp. 411–416.

9. Behr, M., Rosentritt, M., Wimmer, J., Lang, R., Kolbeck, C., Bürgers, R., and Handel, G. (2009), Self-adhesive resin cement versus zinc phosphate luting material: a prospective clinical trial begun 2003, *Dent. Mater.*, **25**, pp. 601–604.

10. Hermansson, L. (2011), Nanostructural chemically bonded Ca-aluminate based biomaterials, in *Biomaterials: Physics and Chemistry*, Ed. R. Pignatello (INTECH, Rijeka).

11. ISO standards.

12. Pameijer, C. H., Jefferies, S., Lööf, J., and Hermansson, L. (2008),

Microleakage evaluation of Xeracem in cemented crowns, *J. Dent. Res.*, **87**(B), p. 3098.

13. Pameijer, C. H., Zmener, O., Garcia-Godoy, F., and Alvarez-Serrano, S. (2009), Sealing of XeraCem, and controls using a bacterial leakage model, *J. Dent. Res.*, **88**(A), p. 3145.

14. Pameijer, C. H., Jefferies, S. R., Loof, J., Wiksell, E., and Hermansson, L. (2008), In vitro and in vivo biocompatibility tests of XeraCem, *J. Dent. Res.*, **87**(B), p. 3097.

15. Lööf, J., Svahn, F., Jarmar, T., Engqvist, H., and Pameijer, C. H. (2008), A comparative study of the bioactivity of three materials for dental applications, *Dent. Mater.*, **24**, pp. 653–659.

16. Pameijer, C. H., Zmener, O., Alvarez Serrano, S., and Garcia-Godoy, F. (2010), Sealing properties of a calcium aluminate luting agent, *Am. J. Dent.*, **23**, pp. 121–124.

17. Pameijer, C. H., Jefferies, S., Lööf, J., and Hermansson, L. (2008), Microleakage evaluation of Xeracem in cemented crowns, *J. Dent. Res.*, **87**(B), p. 3098.

18. Ivar, J. F., and Ryge, G. (1971), *Criteria for the Clinical Evaluation of Dental Restorative Materials*. San Francisco, Government Printing Office, US Public Health Services Publication No. 790–244.

19. Loe, H., and Silness, J. (1963), Periodontal disease in pregnancy. I. Prevalence and severity, *Acta Odontol. Scand.*, **21**, pp. 533–551.

20. Hermansson, L., and Kraft, H. (2003), Chemically bonded ceramics as biomaterials, *Key Eng. Mater.*, **247**, pp. 437–442.

21. Alamo, H. L., Buruiana, R., Schertzer, L., and Boylan, R. J. (1999), A Comparison of MTA, super-EBA, composite and amalgam as root-end filling materials using a bacterial microleakage model, *Int. Endod. J.*, **32**, pp. 197–203.

22. Dentsply Information (2003), *ProRoot MTA* White Brochure B18b.

23. Heini, P. F., and Berlemann, U. (2001), Bone substitutes in Vertebroplasty, *Eur. Spine J.*, **10**, pp. 205–213.

24. Liebermann, I. H., Togawa, D., and Kayanja, M. M. (2005), Vertebroplasty and kyphoplasty: filler materials, *Spine J.*, **5**, pp. 305–316.

25. Bohner, M. (2000), Calciumortophosphates in medicine: from ceramics to calcium phosphate cements, *Injury*, **31**(Suppl 4), pp. 37–47.

26. Nilsson, M. (2003), *Injectable Calcium Sulphates and Calcium Phosphates as Bone Substitutes*, PhD thesis, Lund University.

27. Engqvist, H., Couillard, M., Botton, G. A., Phaneuf, M. P., Axén, N., Ahnfelt, N.-O., and Hermansson, L. (2005), In vivo bioactivity of a novel mineral based based orthopaedic biocement, *Trends Biomater. Artif. Organs*, **19**, pp. 27–32.

28. Jarmar, T., Uhlin, T., Höglund, U., Thomsen, P., Hermansson, L., and Engqvist, H. (2008), Injectable bone cements for vertebroplasty studied in sheep vertebrae with electron microscopy, *Key Eng. Mater.*, **361–363**, pp. 373–376.

29. Engqvist, H., Edlund, S., Gómez-Ortega, G., Lööf, J., and Hermansson L. (2006), In vitro mechanical properties of a calcium silicate based bone void filler, *Key Eng. Mater.*, **309–311**, pp. 829–832.

30. Faris, A., Engqvist, H., Lööf, J., Ottosson, M., and Hermansson, L. (2006), In vitro bioactivity of injectable ceramic orthopaedic cements, *Key Eng. Mater.*, **309–311**, pp. 833–836.

31. Engqvist, H., Persson, T., Lööf, J., Faris, A., and Hermansson, L. (2008), Chemical stability of a novel bioceramic for stabilisation of vertebtal compression, *Trends Biomater., Artif. Organs,* **21**, pp. 98–106.

32. Lööf, J., Faris, A., Hermansson, L., and Engqvist, H. (2008), In vitro Biomechanical testing of two injectable materials for vertebroplasty in different synthetic bone, *Key Eng. Mater.*, **361–363**, pp. 369–372.

33. Breding, K., and Engqvist, H. (2008), Strength and chemical stability due to aging of two bone void filler materials, *Key Eng. Mater.*, **361–363**, pp. 315–318.

34. Hermansson, L. (2011), Nanostructural chemically bonded Ca-aluminate based biomaterials, in *Biomaterials: Physics and Chemistry*, Ed. R. Pignatello (INTECH, Rijeka).

Chapter 13

Classification and Summary of Beneficial Features of Nanostructural Chemically Bonded Bioceramics

This chapter presents summary aspects and typical and beneficial property properties of nanostructural chemically bonded ceramics as biomaterials and how these bioceramics are classified and related to other biomaterials.

13.1 Introduction: A Classification of Biomaterials

Chemically bonded bioceramics (CBBCs) are mainly found within phosphates, silicates, aluminates, and sulphates, as well as combinations of these systems [1, 2]. The forming reaction is in most cases a hydration process, where the solid part, the original powder, reacts with water. More detailed information is given in Chapter 3.

CBBCs are an important but small part of the whole biomaterials field. In Fig. 13.1 is presented an overview of how CBBCs can be classified as biomaterials.

In Table 13.1 are summarised some general properties of biomaterials.

Nanostructural Bioceramics: Advances in Chemically Bonded Ceramics
Leif Hermansson
Copyright © 2015 Pan Stanford Publishing Pte. Ltd.
ISBN 978-981-4463-43-0 (Hardcover), 978-981-4463-44-7 (eBook)
www.panstanford.com

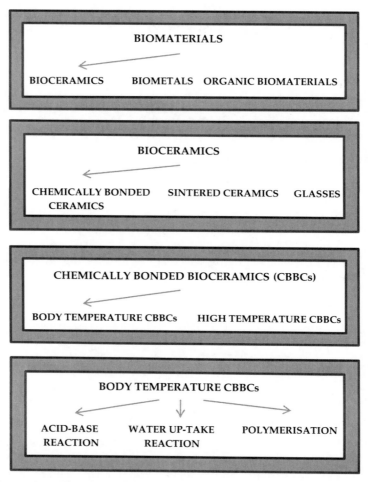

Figure 13.1 Classification of bioceramics.

Nanostructural CBBCs are mainly found within the acid–base-related materials. The base is predominantly found within silicates (C2S and C3S) and aluminates (CA, C12A7, and C3A) with water as the weak acid [3, 4]. For sulphates and phosphates both typical acid–base reactions and a more direct water uptake reaction are possible [5]. Geopolymers can be produced at low temperatures as well as at high temperatures by a combination of different inorganic units in a polymerisation process. At low temperatures amorphous phases appear [6].

Table 13.1 General aspects of some biomaterials

Material aspects	Glass ionomer	Resin (bonded)	Self-adhesive resin	Nanostructural CBBCs
Type of material	Polymer	Monomer	Monomer	Ceramic (polymer)
Hardening mechanism	Cross-linking	Polymerisation	Polymerisation	Acid–base + cross-linking
pH	Acidic	Acidic/neutral	Acidic/neutral	Acidic/basic
Geometrical stability	Non-shrinking	Shrinks	Shrinks	Non-shrinking
Stability over time	Degrades	Degrades	Degrades	Stable
Extra treatment	–	Etching and bonding	–	–
Hydrophilic /hydrophobic	Hydrophilic	Hydrophobic	Initially hydrophilic	Hydrophilic
Integration mechanism	Micromechanical retention, chemical bonding	Adhesion/ micromechanical retention	Adhesion/ micromechanical retention	Nanostructural integration
General behaviour	Irritant	Allergenic	Allergenic	Non-allergenic
Biocompatibility	Good	OK	OK	Good
Bioactivty	No	No	No	Bioactive
Sealing quality	OK	Good but operation-sensitive	OK	Excellent

13.2 Processing and Property Profile

CBBCs can be produced at low temperatures in situ, in vivo. The chemistry of these systems is similar to that of hard tissue in living organisms. CBBCs easily form nanostructures with crystal sizes similar to those found in hard tissue. Both stable and resorbable CBBCs can be produced. The stable phases are found within the $CaO-Al_2O_3-H_2O$ and $CaO-SiO_2-H_2O$ systems, while resorbable phases are seen within the $CaO-P_2O_5-H_2O$ system and within sulphate systems.

CBBCs exhibit several properties suitable for in situ, in vivo placement in hard tissues. This is based on the chemical, physical, and biological features of the biomaterials [7–12]. CBBCs—especially materials based on phosphates, aluminates, and silicates—exhibit a general nanostructure related to both crystals and the porosity between the crystals formed.

Due to a low-solubility product of the phases formed, nanocrystals are easily formed, and it is difficult to avoid nanostructural features. The nanocrystal size is often in the range of 10–40 nm with open porosity with a nanochannel width in the range of 1–3 nm [2, 13]. General properties of CBBCs due to the nanostructures developed deal mainly with:

- high mechanical strength;
- reduced porosity; and
- complete sealing of contact zones to tissue and other biomaterials due to the nanostructural hydration mechanisms.

General properties for CBBCs are presented as [2] follows:

- CBBCs include apatite, the main chemical constituent in hard tissue.
- CBBCs can favourably be produced at body temperature.
- CBBCs tolerate moist conditions.
- Hardening can be controlled to avoid shrinkage.
- The thermal and electrical properties of CBBCs are close to those of hard tissue.

Additives for CBBCs [8, 14] are introduced to promote:

- early and controlled hardening and hydration;
- controlled microstructure and homogeneity, as well as specific properties related to mechanical strength and dimensional stability;

- radio-opacity;
- transparency; and
- biologically related properties, including biocompatibility, bioactivity, controlled resorption, and anti-bacterial properties.

13.3   Unique Properties

Ca-aluminate cements and to some extent also Ca-silicate cements exhibit unique combinations of properties. The specific interesting combination of properties for Ca-aluminate systems is the simultaneous appearing of bioactivity and bacteriostatic and anti-bacterial properties, as well as reduced microleakage.

The anti-bacterial properties of Ca-aluminate and Ca-silicate systems during hydration and as a permanent implant material are due to the following [15, 16]:

- pH during initial hardening—anti-bacterial effect at low and high pH
- Presence of F ions—similarity in size to that of hydroxyl ions
- The nanostructure developed—entrapping of bacteria and growth inhibition
- The surface structure of hydrated Ca-aluminate—fastening of bacteria to the surface and growth inhibition
- Nanoporosity, with nanochannels below 20 nm, often in the range of 1–3 nm, used for controlled slow release of medicaments

The nanochannels surrounding all formed nanosize hydrates will also contribute to pressure relief. Haemacompatibility is concluded for the Ca-aluminate hydrated system [17].

13.4   Applications for Nanostructural Chemically Bonded Bioceramics

Nanostructural CBBCs are based on materials of the $CaO-Al_2O_3-P_2O_5-SiO_2-H_2O$ (CAPSH) system comprising phosphates, aluminates, and silicates. The features and benefits of these systems are summarised in Table 13.2.

Table 13.2   Features and benefits of the CAPSH technology platform

• Nanostructural integration and apatite formation • No shrinkage • Integration/stability/strength • No bonding/dry field required • Variable consistency and compatibility to other materials • Moisture-tolerant	• Reduced risk of secondary caries • No or limited post-operative sensitivity • Longevity/durability • Easy and fast • Broad spectrum of usage with products targeting indication needs • Environmental friendliness

The following product areas for nanostructural CBBCs have been identified on the basis of experimental material data, pre-clinical studies, pilot studies, and ongoing clinical studies [2]:

Dental applications: Dental cements, endodontic products (orthograde and retrograde), sealants, restoratives, underfillings, and pastes for augmentation and dental implant coatings

Orthopaedic applications: Percutaneous vertebraplasty, kyphoplasty, implant coating, and bone void-filling

Drug delivery carrier applications: The chemistry allowing loading of different medicaments within a broad time interval for delivery—from hours to months

Acknowledgement

The author thanks the personnel at Doxa Company, Sweden, and the Materials Science Department at Uppsala University for valuable inputs during a two-decade period.

References

1. Park, J. B. (2008), *Bioceramics* (Springer, New York).

2. Hermansson, L. (2011), Nanostructual chemically bonded Ca-aluminate based biomaterials, in *Biomaterials: Physics and Chemistry*, Ed. R. Pignatello (INTECH, Rijeka).

3. Muan, A, and Osbourne, E. A. (1965), *Phase Equilibria among Oxides*, Ed. (Addison-Wesley, New York).

4. Mangabhai, R. J. (1990), *Calcium Aluminte Cement, Conference Proceedings* (Chapman and Hall).

5. Nilsson, M. (2003), *Injectable Calcium Sulphates and Calcium Phosphates as Bone Substitutes*, PhD thesis, Lund University.

6. Forsgren, J. (2010), PhD thesis, Uppsala University (Acta Universitatis Upsaliensis).

7. Hermansson, L., Lööf, J., and Jarmar, T. (2009), Integration mechanisms towards hard tissue of Ca-aluminate based biomaterials, *Key Eng. Mater.*, **396–398**, pp. 183–186.

8. Kraft, L. (2002), *Calcium Aluminate Based Cement as Dental Restorative Materials*, PhD thesis, Uppsala University.

9. Hermansson, L., Kraft, L., Lindqvist, K., Ahnfelt, N.-O., and Engqvist, H. (2008), Flexural strength measurement of ceramic dental restorative materials, *Key Eng. Mater.*, **361–363**, pp. 873–876.

10. Engqvist, H., Edlund, S., Gómez-Ortega, G., Lööf, J., and Hermansson L. (2006), In vitro mechanical properties of a calcium silicate based bone void filler, *Key Eng. Mater.*, **309–311**, pp. 829–832.

11. Lööf, J., Engqvist, H., Gómez-Ortega, G., Spengler, H., Ahnfelt, N.-O., and Hermansson, L. (2005), Mechanical property aspects of a biomineral based dental restorative system, *Key Eng. Mater.*, **284–286,** pp. 741–744.

12. Lööf, J., Engqvist, H., Hermansson, L., and Ahnfelt, N.-O. (2004), Mechanical testing of chemically bonded bioactive ceramic materials, *Key Eng. Mater.*, **254–256,** pp. 51–547.

13. Powers, T. C., and Brownyard, T. L. (1946), Studies of the physical properties of hardened cement paste, Publ. of Res. Lab. of the Portland Cement association, Chicago, USA.

14. Lööf, J. (2002), *Calcium-Aluminate as Biomaterial: Synthesis, Design and Evaluation*, PhD thesis, Uppsala University.

15. Hermansson, L. (2012), Aspects of antibacterial properties of nanostructural calcium aluminate based biomaterials, *Ceramic. Eng. Sci. Proc.*, **33**, pp. 57–64.

16. Unosson, E., Cai, E., Jiang, J., Lööf, J., and Engqvist, H. (2012), Antibacterial properties of dental luting agents: potential to hinder the development of secondary caries. *Int. J. Dentistry*, **2012** ID 529495.

17. Axen, N., Ahnfelt N.-O., Persson, T., Hermansson, L., Sanchez, J., and Larsson R. (2005), A comparative evaluation of orthopaedic cements in human whole blood, in *Advances in Bioceramics and Biocomposites, Ceramic Engineering and Science Proceedings*, Vol. **26**(6) (ed. M. Mizuno), John Wiley & Sons, Inc., Hoboken)

Chapter 14

Future Aspects of Nanostructural Chemically Bonded Bioceramics

This chapter discusses the potential evolution of nanostructural chemically bonded ceramics as biomaterials in the coming decades.

14.1 Introduction

Chemically bonded bioceramics (CBBCs) have now passed the technology platform, that is, how to develop materials with specific properties, and entered the product platform, that is, how to make the biomaterials practical and safe for specific applications. This is discussed in some detail in Chapter 13. In this chapter the focus will be on possible development to enlarge the product platform and to improve the safety and biological interference with other biomaterials and biological tissue.

Some of the CBBCs are relatively new as biomaterials, but CBBCs have been established as a complementary bioceramic group [1–4]. The most specific features are the in situ, in vivo formation of the material to fill hard tissue voids, within both dentistry and orthopaedics [5].

The presentation below is an attempt to enlighten areas where possible future activities and achievements will be seen.

Nanostructural Bioceramics: Advances in Chemically Bonded Ceramics
Leif Hermansson
Copyright © 2015 Pan Stanford Publishing Pte. Ltd.
ISBN 978-981-4463-43-0 (Hardcover), 978-981-4463-44-7 (eBook)
www.panstanford.com

14.2 Possible Future Developments

The discussion will be structured related to the following topics: nanostructural CBBC materials, specific properties, the future product platform, active additives, drug delivery, and third-generation biomaterials.

14.2.1 Nanostructural CBBC Materials

Within specifically the calcium aluminate hydrate (CAH) and calcium silicate hydrate (CSH) and combinations of CAH and CSH with Ca-phosphates (CPs) new improved bioceramics for specific applications are foreseen [6, 7]. Active additional binding systems are also foreseen.

Another field where development is expected deals with raw material production. Still there are only few raw materials that could be used directly. In much companies rely on raw material synthesis.

14.2.2 Specific Properties

A need within CBBC materials relates to improve safety in the mechanical properties of the end products [8]. Important areas here are handling aspects to avoid introduced defects and a need to improve fracture toughness.

Another materials aspect is the need of a more aesthetic appearance of the CBBCs [9]. This relates to dentistry. The transparency needs to be approximately as that of enamel.

14.2.3 Active Additives

Due to the almost perfect nanoporosity structure of several CBBCs, medically active additives and health-related additives can easily be incorporated in the material. This is an area of high interest.

Developments of a combination of general biomaterials and at the same time carriers for drug delivery are foreseen [10]. The antibacterial properties of some of the CBBCs should also be taken into account [11].

14.2.4 Third-Generation Biomaterials

For even more biologically related nanostructural bioceramics stem cell–produced materials are expected. However, a restriction is the time before cells in the biomaterial can be used. That's why a combination of nanostructural CBBCs and stem cell activity is likely in the development of these third-generation biomaterials, especially if load-bearing properties are required.

14.3 Conclusion

To be able to compete with established materials—primarily polymer-based biomaterials—more products have to reach users of biomaterials. An extended product platform for nanostructutal CBBCs is foreseen.

Within dentistry the following product platform for nanostructural CBBCs is of great importance:

Dental cements:

      Next generation

      Temporary and semi-permanent cements

Endodontics:

      New CBBCs as

          Orthograde fillings

          Retrograde fillings

Dental fillings:

      Paedodontic fillings

      General restoratives

      Fissure sealings

As a significant number of dental restorations today are replacements of old, failed tooth fillings, it is clear that tackling this problem is of great concern and a market need [12].

Within orthopaedics the following product platform for nanostructural CBBCs is foreseen:

Vertebroplasty:

      Percutaneous vertebroplasty (PVP)

      Kypho-vertebroplasty (KVP)

Bone void fillers:

General

Craniofacial

For both dental and orthopaedic implants a development of new coatings is foreseen [13]. This is related to the concern of perimplantitus.

Acknowledgement

The author thanks the personnel at Doxa Company, Sweden, and the Materials Science Department at Uppsala University for valuable inputs during a two-decade period.

References

1. Park, J. B., and Lakes, R. S. (2007), *Biomaterials: An Introduction* (Springer, New York).

2. Park, J. B. (2008), *Bioceramics* (Springer, New York).

3. Duxson, P., Fernández-Jiménez, A., Provis, J. L., Lukey, G. C., Palomo, A., van Deventer, J. S. J. (2007), Geopolymer technology: the current state of the art, *J. Mater. Sci.*, **42**(9), pp. 2917–2933.

4. Hench, L. (1998), Biomaterials: a forecast for the future, *Biomaterials*, **19**, pp. 1419–1423.

5. Hermansson, L. (2011), Nanostructural chemically bonded Ca-aluminate based biomaterials, in *Biomaterials: Physics and Chemistry*, Ed. R. Pignatello (INTECH, Rijeka), pp. 831–834.

6. Hermansson, L., Lööf, J., Jarmar, T. (2009), Integration mechanisms towards hard tissue of Ca-aluminate based biomaterials, *Key Eng. Mater.*, **396–398**, pp. 183–186.

7. Engqvist, H., Schultz-Walz, J. E., Lööf, J., Botton, G. A., Mayer, D., Phaneuf, M. W., Ahnfelt, N. O., Hermansson, L. (2004), Chemical and biological integration of a mouldable bioactive ceramic material capable of forming apatite in vivo in teeth, *Biomaterials*, **25**, pp. 2781–2787.

8. Kraft, L. (2002), *Calcium Aluminate Based Cement as Dental Restorative Materials*, PhD thesis, Uppsala University.

9. Engqvist, H., Lööf, J., Uppström, S., Phaneuf, M. W., Jonsson, J. C., Hermansson, L., Ahnfeldt, N-O. (2004), Transmittance of a bioceramic calcium aluminate based dental restorative material, *J. Biomed. Mater. Res. Part B: Appl. Biomater.*, **69**(1), pp. 94–98.

10. Hermansson, L. (2010), Chemically bonded bioceramic carrier systems for drug delivery, *Ceram. Eng. Sci. Proc.,* **3**, pp. 77–88.

11. Hermansson, L. (2012), Aspects of antibacterial properties of nanostructural calcium aluminate based biomaterials, *Ceramic. Eng. Sci. Proc.,* **33**, pp. 57–64.

12. Mjör, I. A, Moorhead, J. E and Dahl, J. E. (2000), *Int. Den. J.,* **50**(6), pp. 50.

13. Axén, N., Engqvist, H., Lööf, J., Thomsen P., Hermansson, L. (2005), In vivo hydrating calcium aluminate coatings for anchoring of metal implants in bone, *Key Eng. Mater.,* **284–286,** pp. 831–834.

Definitions

Allograft: Transplant between unrelated individuals of the same species

Autograft: Transplant within an individual, from one part of the body to another

Bioactivity: A material which elicits a specific response at the interface of the material, which results in the formation of a bond between the tissue and the material

Biocompatibility: The ability of a material to perform with an appropriate host response in a specific application; tissue friendliness in a specific application

Biodegradation: Gradual breakdown of a material by bacterial or enzymatic action

Biomaterial: A material intended to interface with biological systems to evaluate, treat, augment, or replace any tissue, organ, or function of the body (1992); a nonviable material used in a medical device intended to interact Williams et al biological systems

Biomimics: Materials science and engineering through biology

Buccal: The tooth surface which is next to the cheek

Cement abbreviation system: A for $Al2O_3$, C for CaO, S for SiO_2, H for H_2O, and so on

Centrals: The two upper and two lower teeth in the very centre of the mouth

Graft: A transplant

Host response: The reaction of a living system to the presence of a material

Implant: A medical device made from one or more biomaterials which is intentionally placed within the body

Incisal: The biting edge of the centrals and laterals

in vitro: An experiment carried out in a controlled environment outside the living organism

in vivo: An experiment carried out in or on a living organism

Laterals: Teeth just adjacent to the centrals (see centrals)

Lingual: The tooth surface next to the tongue

Nano: 10^{-9} m

Nanosize: Size interval 1 – 100 nm

Occlusal: The chewing or grinding surface of posterior teeth (molars and pre-molars)

Osteoconduction: A material property which allows only an extracellular response at the interface

Osteoinduction: Both intracellular and extracellular affinity of bone formation

Proximal: The surfaces of teeth which touch the next teeth

Resorption: Absorption of a material, a polymer or a ceramic

Stem cells: Undifferentiated cells capable of proliferation, self-renewal, and differentiation into at least one type of specialised cell

Tissue engineering: An interdisciplinary field for the development of biological substitutes containing living cells

Transplant: A complete structure, such as an organ, which is transferred from a site in a donor to a site in a recipient for the purpose of reconstruction of the recipient site

Abbreviations

A	Al_2O_3
AH_3	$Al(OH)_3$ or $Al_2O_3 \cdot H_2O$
BET	Brunauer–Emmett–Teller test for specific surface area
BMP	bone morphogenetic protein
C	CaO
CA	calcium mono-aluminate, $CaO \cdot Al_2O_3$
CA_2	calcium di-aluminate, $CaO \cdot 2Al_2O_3$
C_3A	$3CaO \cdot Al_2O_3$
$C_{12}A_7$	$12CaO \cdot 7Al_2O_3$
CAC	calcium aluminate cement
CAH	calcium aluminate hydrate
CAPH	$CaO–Al_2O_3–P_2O_5–H_2O$, calcium aluminate–calcium phosphate
CAPSH	$CaO–Al_2O_3–P_2O_5–SiO_2–H_2O$
CBC	chemically bonded ceramic
CBBC	chemically bonded bioceramic
CitA	Citric acid
CIP	Control Investigation Plan
CP	Ca-phosphate
CPC	calcium phosphate cement
CPH	$CaO–P_2O_5–H_2O$
cpo	chronic perapical osteitis
CRO	Contract Research Organisation
CS	$CaO \cdot SiO_2$
C_2S	$CaO \cdot 2SiO_2$
C_3S	$CaO \cdot 3SiO_2$
CSC	calcium silicate cement

CSH	calcium silicate hydrate
CT	computed tomography
EDAX	energy-dispersive X-ray analysis
EDS	energy-dispersive X-ray spectrometer
ESEM	environmental scanning electron microscopy
FIB	focused ion beam
FM	filler material
FPD	fixed partial denture
FT-IR	Fourier transform infrared spectroscopy
GCP	good clinical practice
GIC	glass ionomer cement (polyalkenoate)
H	H_2O
HA	hydroxyapatite
HRTEM	high-resolution transmission electron microscopy
HV	Vickers hardness
l/p	liquid-to-powder ratio
IBT	inflatable balloon tamp
IPA	isopropanole
ISO	International Organisation for Standardisation
KVP	kypho-vertebroplasty
MODDE	software for design of experiments and optimisation
MRI	magnetic resonance imaging
μ-SiO_2	fumed silica
Na-PAA	poly(acrylic acid) sodium salt
Na-PAMA	poly(acrylic-co-maleic acid) sodium salt
OPC	ordinary Portland cement
PAA	poly(acrylic acid)
PAMA	poly(acrylic-co-maleic cid)
PBS	phosphate buffered saline
PI	principal investigator, who leads the study conduct at an individual study site (every study site has a PI.)
PIC	patient informed consent
PLS	partial least square
PMMA	poly(methyl methacrylate)

PVC	poly(vinyl chloride)
PVP	percutaneous vertebroplasty
QoL	quality of life
rf	root filling
RMGI	resin-modified glass ionomer
SAE	serious adverse event
SBF	simulated body fluid
SEM	scanning electron microscopy
ST	setting time
STEM	scanning transmission electron microscopy
TAT	thrombin–anti-thrombin complex
TCP	tri-calcium phosphate
TEM	transmission electron microscopy
TGB-β	transforming growth factor beta
TSH	thyroid-stimulating hormone
TTA	tartaric acid
USPHS	United States Public Health Service
VAS	visual analogue scale
VCP	vertebral compression fracture
w/c	water-to-cement ratio
w/p	water-to-powder ratio
WT	working time
XRD	X-ray diffraction
XPS	X-ray photoelectron spectroscopy

Index